PUBLISHERS' NOTE

As its title implies, the Series in which this volume appears has two purposes. One is to encourage the publication of monographs on advanced or specialised topics in, or related to, the theory and applications of probability and statistics; such works may sometimes be more suited to the present form of publication because the topic may not have reached the stage where a comprehensive treatment is desirable. The second purpose is to make available to a wider public concise courses in the field of probability and statistics, which are sometimes based on unpublished lectures.

The series was edited from its inception in 1957 by Sir Maurice Kendall, under whose editorship the first 21 volumes in the Series appeared. He was succeeded as editor in 1965 by Professor Alan Stuart.

The publishers will be interested in approaches from any authors who have work of importance suitable for the Series.

CHARLES GRIFFIN & CO. LTD.

TITLES ISSUED IN GRIFFIN'S STATISTICAL MONOGRAPH SERIES

No. 1: *The analysis of multiple time-series* — M. H. QUENOUILLE
No. 2: *A course in multivariate analysis* — SIR MAURICE KENDALL
No. 3: *The fundamentals of statistical reasoning* — M. H. QUENOUILLE
No. 4: *Basic ideas of scientific sampling* — A. STUART
No. 6: *An introduction to infinitely many variates* — E. A. ROBINSON
No. 7: *Mathematical methods in the theory of queueing* — A. Y. KHINTCHINE
No. 8: *A course in the geometry of n dimensions* — SIR MAURICE KENDALL
No. 9: *Random wavelets and cybernetic systems* — E. A. ROBINSON
No. 10: *Geometrical probability* — SIR MAURICE KENDALL and P. A. P. MORAN
No. 11: *An introduction to symbolic programming* — P. WEGNER
No. 12: *The method of paired comparisons* — H. A. DAVID
No. 13: *Statistical assessment of the life characteristic: a bibliographic guide* — W. R. BUCKLAND
No. 14: *Applications of characteristic functions* — E. LUKACS and R. G. LAHA
No. 15: *Elements of linear programming with economic applications* — R. C. GEARY and J. E. SPENCER
No. 16: *Inequalities on distribution functions* — H. J. GODWIN
No. 17: *Green's function methods in probability theory* — J. KEILSON
No. 18: *The analysis of variance: a basic course* — A. HUITSON
No. 19: *The linear hypothesis: a general theory* — G. A. F. SEBER
No. 20: *Econometric techniques and problems* — C. E. V. LESER
No. 21: *Stochastically dependent equations: an introductory text for econometricians* — P. R. FISK
No. 22: *Patterns and configurations in finite spaces* — S. VAJDA
No. 23: *The mathematics of experimental design: incomplete block designs and Latin squares* — S. VAJDA
No. 24: *Cumulative sum tests: theory and practice* — C. S. VAN DOBBEN DE BRUYN
No. 25: *Statistical models and their experimental application* — P. OTTESTAD
No. 26: *Statistical tolerance regions: classical and Bayesian* — I. GUTTMAN
No. 27: *Families of bivariate distributions* — K. V. MARDIA
No. 28: *Generalized inverse matrices with applications to statistics* — R. M. PRINGLE and A. A. RAYNER
No. 29: *The generation of random variates* — T. G. NEWMAN and P. L. ODELL
No. 30: *Families of frequency distributions* — J. K. ORD
No. 31: *A first course in statistics* — F. N. DAVID
No. 32: *The logit transformation, with special reference to its uses in bioassay* — W. D. ASHTON
No. 33: *Regression estimation from grouped observations* — Y. HAITOVSKY
No. 34: *Stochastic point processes and their applications* — S. K. SRINIVASAN
No. 35: *The analysis of categorical data* — R. L. PLACKETT
No. 36: *Input-output analysis and its applications* — R. O'CONNOR and E. W. HENRY
No. 37: *Sampling inspection in statistical quality control* — W. C. GUENTHER
No. 38: *Maximum likelihood estimation in small samples* — L. R. SHENTON and K. O. BOWMAN

No. 5, formerly *Characteristic Functions* by E. Lukacs, is now published independently of the Series

Maximum Likelihood Estimation in Small Samples

L. R. SHENTON
University of Georgia

K. O. BOWMAN
Union Carbide Corporation, Nuclear Division
Oak Ridge, Tennessee

Monograph No. 38
Series Editor
ALAN STUART

CHARLES GRIFFIN & COMPANY LTD
London and High Wycombe

CHARLES GRIFFIN & COMPANY LIMITED
Registered Office: 5A Crendon Street, High Wycombe HP13 6LE

Copyright © L. R. Shenton & K. O. Bowman, 1977

All rights reserved. No part of this book may be reproduced or transmitted in any form or by any means, electronic or mechanical, including photocopying, recording, or by any information storage and retrieval system, without permission in writing from the Publisher.

First published   1977

ISBN: 0 85264 238 5

Printed in Great Britain at the Alden Press
Oxford London and Northampton

CONTENTS

PREFACE — page ix

1 OUTLINES OF BASIC THEORY — 1

  1.1 Introduction — 1
  1.2 Standard maximum likelihood theory — 1
    1.2.1 Likelihood function — 1
    1.2.2 Consistency — 3
    1.2.3 Bias — 5
    1.2.4 Asymptotic normality — 6
    1.2.5 MLE as efficient estimators — 7
    1.2.6 Multiparameter estimation — 9
    1.2.7 Sufficiency — 12
  1.3 Small sample problems — 12
    1.3.1 Introduction — 12
    1.3.2 Illustrations — 13
    1.3.3 — 22
    1.3.4 Summary of types of estimation problems — 22
    1.3.5 Asymptotic series — 23
    1.3.6 Asymptotic moments — 25
    1.3.7 Illustrations of asymptotic expansions in statistics — 26
    1.3.8 Asymptotics in general — 30
  References — 31

2 SINGLE PARAMETER ESTIMATION — 34

  2.1 The problem — 34
  2.2 Methods available for moment evaluation — 34
  2.3 Expectation of products of linear forms — 35
  2.4 Order of magnitude of products of linear forms — 36
  2.5 Lagrange's expansion — 36
  2.6 Adjusted order of magnitude method — 38
  2.7 c-variate Taylor approach — 41
    2.7.1 — 41
    2.7.2 Derivatives of $\hat{\theta}$ — 41
    2.7.3 Contributing terms in the sampling moments — 42
    2.7.4 Detailed consideration of the bias of $\hat{\theta}$ — 43
    2.7.5 — 44
    2.7.6 First four moments using logarithmic derivatives — 44
    2.7.7 Skewness and kurtosis — 49
  2.8 Moments using derivatives of the probability function — 51
  2.9 Illustrations — 56
    2.9.1 Probability function linear in θ — 56
    2.9.2 Normal distribution with mean θ and unit variance — 57
    2.9.3 Normal distribution with zero mean and variance θ — 58
    2.9.4 Poisson distribution with mean θ — 59
    2.9.5 Fisher's "Nile" problem — 60
  References — 62

3 BIAS AND COVARIANCE IN MULTIPARAMETER ESTIMATION — 63

    3.1 Introduction — 63
    3.2 Notation — 63
    3.3 Taylor expansion — 64
    3.4 Covariances — 66
    3.5 Bias — 68
    3.6 Special cases and the covariance — 71
        3.6.1 Case when the parameters are equal — 71
        3.6.2 Orthogonal parameters — 72
        3.6.3 Two parameters with $\hat{\theta}_2$ the sample mean — 73
        3.6.4 Estimators for the negative binomial distribution — 74
    3.7 Special cases and the bias — 75
    3.8 Adjusted order of magnitude method — 77
    References — 77

4 BIASES AND COVARIANCES FOR ESTIMATORS IN NON-REGULAR CASES — 79

    4.1 Introduction — 79
    4.2 Negative binomial distribution — 80
    References — 90
    4.3 Neyman Type A distribution — 91
    References — 93
    4.4 Pólya–Aeppli distribution — 108
    References — 110
    4.5 Three-parameter Gamma distribution — 110
    References — 113
    4.6 Two-parameter Hermite — 113
    References — 120
    4.7 An exponential regression model — 121

5 SPECIAL DENSITY ESTIMATION — 126

    5.1 Introduction — 126
    5.2 Logarithmic series distribution — 126
        5.2.1 — 126
        5.2.2 Basic formulae — 126
        5.2.3 Central moments — 127
        5.2.4 Derivatives of $\hat{\theta}$ — 129
        5.2.5 Comments on the moments — 130
        5.2.6 Skewness and kurtosis — 136
        5.2.7 Verification of the results — 137
        5.2.8 Conclusion and discussion — 142
    References — 143
    5.3 Zero-truncated binomial distribution — 143
        5.3.1 Basic formulae — 143
        5.3.2 Taylor expansion — 143
        5.3.3 Tabulations — 144
        5.3.4 Comments on the tables — 144
    References — 149
    5.4 Two-parameter Gamma distribution — 149
        5.4.1 Introduction — 149
        5.4.2 Basic formulae — 150
        5.4.3 Expansions for the ml estimators, N large — 150
        5.4.4 Expansions for ρ large — 152
        5.4.5 Validation of the moments — 154

	5.4.6 Concluding remarks	154
	References	156
6	SUMMARY AND CONCLUSIONS	158
	6.1	158
	6.2 The method of moments	158
	6.3 Moment estimators related to the likelihood function	158
	6.4 Mixed moment estimators	161
	6.5 Estimators derived from measures of discrepancy	161
	6.6 Moment estimators and numerical evaluation	162
	6.7 Conclusions	163
	References	165

APPENDIX A A factorial series (in N) for E $(\hat{\theta} - \theta)^s$ 166

    A.1 Introduction   166
    A.2 Restrictions on the partitions   167
    A.3 The general case   172
    A.4 Factorial series for bias and variance up to N^{-3} 174
    References   176

APPENDIX B Summation of asymptotic series   177

    B.1 Introduction   177
    B.2 Summation algorithms   177
    B.3 Illustrations   180
    B.4 Remarks   181
    References   181

INDEX   183

AUTHORS CITED   185

PREFACE

Most of this monograph is the result of work carried out over the last dozen years or so by students, colleagues and the two authors. It began earlier for one of us who was baffled by the terminology, current at the time, concerning asymptotic variance and covariance, asymptotic efficiency and the like; expressions relating to "information in the sample", "sufficient estimators", etc., were confusing, although the definitions were doubtless intended to have the opposite effect. There was a great outpouring of theoretical articles, and indeed many applied studies as well.

The first attempt at evaluating higher-order terms for maximum likelihood estimators of a single parameter was carried out during 1962–4 by the authors at Virginia Polytechnic Institute; we are grateful to Boyd Harshbarger for his interest at this time. Small (by present standards) electronic computers were appearing at most colleges, and the evaluation of the first set of higher-order terms accounted for a hundred hours or more on IBM 1620. The multi-parameter biases and covariances to order N^{-2} seemed almost beyond hope, and only special cases (simultaneous estimation of two parameters when the sample mean is one estimator) were carried out. In 1964–5 the problem was taken up at Oak Ridge National Laboratory, where larger and faster machines were available. It was here that general formulae for N^{-2} biases and covariances were finally settled; since very few examples could be worked through algebraically, because of the considerable complexity, the formulae lay in limbo, although their programming was slowly taking shape.

Another step forward was effected by Danny Sheehan working at the Computer Center, University of Georgia (1966–8); Sheehan worked out evaluations for terms up to order N^{-6} (a few up to N^{-9}) for estimators of the two parameters of the negative binomial distribution.

In the last few years we have taken great pains to assure ourselves that the formulae are correct, and to test out in various ways numerical output. Most of the work has been done on IBM System 360 Model 91 at Oak Ridge National Laboratory, with support work on the CDC machine at Georgia University.

A number of colleagues and co-workers have assisted in one way or another; we mention Professor Ray Myers, Shiras Elliott, Dr Y. C. Patel, Dr Tom Mason, and in particular Dr Danny Sheehan.

Deciding on the format of the monograph was not an easy matter, since the correct setting out and explanation of the rather dull basic formulae was of paramount importance. Again, it is difficult to judge what the average statistics student knows (or should know) about basic asymptotic theory. In the end, we have sketched out basic maximum likelihood theory (in so far as it is needed to understand the monograph) and basic asymptotology.

As for statistical knowledge, readers could not do better than use as a constant companion Kendall and Stuart's *Advanced Theory of Statistics* (especially Vol. 2) and Kendall's *Exercises in Theoretical Statistics*.

Since completing the manuscript some thought has been given to the possibility of evaluating higher-order terms; for example, third-order terms in the variances in simultaneous estimation. A summary of this incomplete study is given in Appendix A. Moreover, in recent times, there is a good deal of interest and progress in summation techniques for divergent series. It therefore seemed appropriate to mention in Appendix B some aspects of this subject that are relevant to the theme of the present work.

We gratefully acknowledge permission to reproduce material from our articles in (i) *Biometrika*, granted by Biometrika Trustees, (ii) *Journal of Statistical Computation and Simulation*, granted by Dr Richard Krutchkoff on behalf of Gordon and Breach, Science Publishers, London, (iii) *Journal of the Royal Statistical Society*, granted by the Editors, (iv) *Technometrics* and *Journal of the American Statistical Association*, granted by the Managing Director of A.S.A., and (v) *Sankhyā*, granted by the Editors. We are especially indebted to Pennsylvania State University Press, University Park, Pennsylvania, for permission to reprint tables and charts from pages 127–50 of *Random Counts in Models and Structures,* Vol. 1 in the series *Random Counts in Scientific Work*, ed. G. P. Patil.

We are indebted to Dr J. G. Skellam, Professor Jim Douglas and Professor C. D. Kemp for criticism of an earlier draft. Also we would like to mention the cooperation of Dr James Carmon, Director of the Office of Computing Activities, University of Georgia, and Dr D. A. Gardiner and staff at Computer Science Division, Union Carbide Corporation Nuclear Division. For a detailed criticism of the manuscript and a number of valuable constructive comments we are indebted to the Editor of the series, Professor Alan Stuart. No small part of the credit for producing a lucid account of the subject, in so far as that has been achieved, is due to Professor Stuart's patient handling of the manuscript.

Miss Becky Cape deserves commendation for her excellent typing, and for her patience and forbearance in keeping up with the numerous modifications and revisions.

Lastly, Mr E. V. Burke of the Editorial Department of Charles Griffin and Company Limited has made many valuable suggestions regarding consistency of notation, etc., which we greatly appreciate.

L.R.S.
K.O.B.

1 OUTLINES OF BASIC THEORY

1.1 Introduction

The intention of this monograph is to set out a methodology developed by the authors over the last dozen years or so, for dealing with maximum likelihood estimators[*] (mle) in small samples. The methods involve a mixture of numerical analysis and classical asymptotic theory, and assume that in the final analysis a digital computer will be used.

It is beyond our scope to state precisely and prove rigorously the varied and intricate properties of mle, which from a modern viewpoint require relatively advanced mathematics (a recent survey of the subject as a whole is given by Norden (1972, 1973)). However, we have thought it necessary to give a brief summary of the main properties which should be kept in mind, and this summary is couched in language which it is hoped will be readily understood by (i) anyone who has had an introduction to estimation theory, and (ii) applied researchers in ecology, biology, engineering, and other physical sciences. Proofs and fuller accounts will be found in Kendall and Stuart (1969, 1973), Cramér (1946), Feller (1966), Rao (1952, 1965). Historical aspects of the subject are given in Rao (1962) and in particular, historical aspects of likelihood are discussed in Edwards (1974). An account of the main aspects of the theory and application of estimation with a careful consideration of asymptotic theory is given in Cox and Hinkley (1974).

1.2 Standard maximum likelihood theory

1.2.1 Likelihood function

For a set of independent and identically distributed observations $x_1, x_2, \ldots, x_n$ depending on a parameter θ, the Likelihood Function[†] *(LF) is*

$$L(x|\theta) = \prod_{j=1}^{n} f(x_j|\theta), \qquad (1.1)$$

where $f(\cdot|\theta)$ refers to the density of a single observation.

---

[*] We shall use "mle" to refer to maximum likelihood estimation (estimator) according to the context. The distinction between an estimate and an estimator is taken for granted.

[†] The likelihood function is related to the probability of the sample in the discrete case, and to the joint density of the sample in the continuous case. It is not always possible to encompass this dichotomy in every aspect of mle.

Fisher (1921) introduced the concept of estimating θ from the observations by finding $\hat{\theta}$ such that for $\theta \in \Theta$ (the parameter space)

$$L(x|\hat{\theta}) \geq L(x|\theta). \tag{1.2}$$

There are at least two possibilities. If Θ is a discrete set, then search $L(x|\theta)$ for the θ which makes it largest. If $L(x|\theta)$ is continuous for θ in some interval, then zero values of the derivative of the likelihood in general locate stationary values; *i.e., we solve*

$$\frac{\partial \ln L}{\partial \theta} = 0$$

for admissible values of θ (i.e., $\theta \in \Theta$). It is possible for more than one solution to exist.

It is interesting to note that if $t(\theta)$ is a differentiable function of θ, then

$$\frac{\partial \ln L}{\partial t(\theta)} = \frac{\partial \ln L}{\partial \theta} \frac{\partial t(\theta)}{\partial \theta},$$

so that if $\hat{\theta}$ is a mle of θ, then $t(\hat{\theta})$ is a mle of $t(\theta)$. For example, the mle of μ^2 in sampling from $N(\mu, 1)$ is $\bar{x}^2$. This property is perhaps not particularly useful, for it does not imply, for example, that

$$E\{t(\hat{\theta})\} = t\{E(\hat{\theta})\},$$

and indeed this will in general be true only if **t** is a linear function of $\hat{\theta}$.

The generalization to simultaneous maximum-likelihood estimators is clear; we seek solutions to (1.2) where $\theta = (\theta_1, \theta_2, \ldots, \theta_h)$. In practice we consider solving

$$\frac{\partial \ln L}{\partial \theta_j} = 0, \quad j = 1, 2, \ldots, h, \theta \in \Theta.$$

Example 1.1 Consider a population of categorized data, the probability of an occurrence in the jth class being $p_j(\theta)$, with n_j observations for a sample of N.

	Categories						Σ_j
	1	2	...	j	...	w	
Probability	p_1	p_2	...	p_j	...	p_w	1
Observed	n_1	n_2	...	n_j	...	n_w	N

The likelihood function is now

$$L(\theta) = k_n \prod_{j=1}^{w} p_j^{n_j}$$

and the mle is a solution to

$$\sum \frac{n_j}{p_j} \frac{\partial p_j}{\partial \theta} = 0,$$

the differentiability of the p's for $\theta \in \Theta$ being assumed. Find the mle for θ for the case

$$p_1 = \tfrac{1}{4}(2+\theta), \quad p_2 = p_3 = \tfrac{1}{4}(1-\theta), \quad p_4 = \tfrac{1}{4}\theta \quad (0 \leqslant \theta \leqslant 1)$$

(see Haldane, 1953; Haldane and Smith, 1956; Fisher, 1925-; Kendall and Stuart, vol. 2, 52-4; also *1.3.7(b), below*).

Example 1.2 A regression model takes the form

$$y_j = a\, e^{b(x_j - \bar{x})} \epsilon_j \quad (j = 1, 2, \ldots, n),$$

where $\bar{x} = \sum x_j/n$, and $(\epsilon_1, \epsilon_2, \ldots, \epsilon_n)$ are independently and identically distributed exponential random variables with unit mean. Thus

$$f(\epsilon_j) = e^{-\epsilon_j}, \quad \epsilon_j > 0$$
$$= 0, \quad \text{otherwise.}$$

Show that the mle of a and b satisfy

$$\sum y_j d_j \exp(-\hat{b} d_j) = 0$$
$$\sum y_j \exp(-\hat{b} d_j) = n\hat{a},$$

where $d_j = x_j - \bar{x}$. (See *4.7.1*.)

1.2.2 Consistency

An estimator $\hat{\theta}_n(x_1, x_2, \ldots, x_n)$ is consistent for θ if

$$p_r\{|\hat{\theta}_n - \theta| < \epsilon\} > 1 - \eta$$

for any small $\epsilon > 0$, $\eta > 0$ and $\eta > N$, $\theta \in \Theta$.

A preferred form is

$$p_r\{|\hat{\theta}_n - \theta| > \epsilon\} \to 0 \quad \text{as } n \to \infty,$$

or $\hat{\theta}_n \xrightarrow{P} \theta$ for $\theta \in \Theta$. This means that for large samples, consistent estimators zero-in on the true value of the parameter, not converging in a mathematical sense (that is, with certainty), but in long sequences of trials, nearly always settling down to the parameter value θ.

The form of the condition depends on knowing the distribution of $\hat{\theta}_n$, and in addition being able to manipulate mathematically an absolute deviation. However, the Bienaymé–Chebyshev inequality in many cases resolves this problem (Kendall and Stuart, vol. 1, p. 88; vol. 2, pp. 3-4; Godwin (1964)). For if the random variable X has $EX^2 < \infty$, $EX = \mu$, $\text{Var } X = \sigma^2$, then

$$P_r\{|X-\mu|<t\} \geq 1-\sigma^2/t, \quad t>0.$$

Assume then that
$$E(\hat{\theta}_n) = \theta + \theta_1(n),$$
$$\text{Var } \hat{\theta}_n = \sigma^2(n),$$

where $\theta_1, \sigma^2 \to 0$ as $n \to \infty$. Then
$$P_r\{|X-\theta-\theta_1(n)|<\epsilon\} \geq 1-\sigma^2(n)/\epsilon,$$

so that for $\epsilon > 0$ and small, the probability tends to unity.

Example 1.3 The mle of μ and v in sampling from the normal $N(\mu, v)$ are
$$\hat{\mu} = \bar{x}, \quad \hat{v} = \Sigma(x-\bar{x})^2/n,$$

where $\bar{x}$ is the sample mean. Show that
$$E(\hat{\mu}) = \mu, \quad E(\hat{v}) = (1-1/n)v, \quad \text{and}$$
$$\text{Var } \hat{\mu} = v/n, \quad \text{Var } \hat{v} = 2(n-1)v^2/n^2;$$

also show that $\hat{\mu}$ tends in probability to μ as $n \to \infty$; similarly for $\hat{v}$.

Example 1.4 Show that the mle of θ in sampling from $N(\theta, \theta)$ (where θ belongs to the positive reals) is given by
$$2\hat{\theta} = -1 + \sqrt{(1+4m'_2)},$$

where $m'_2 = \Sigma x_j^2/n$ is the second moment of the sample.

Prove that the variance of $\hat{\theta}$ in terms of the mean ($v'_1 = E\hat{\theta}$) is given by
$$\text{Var } \hat{\theta} = (\theta - v'_1)(1 + \theta + v'_1).$$

Deduce that $E(\hat{\theta}) \neq \theta$. Prove also for derivatives of the likelihood, that
$$E \frac{\partial^s \ln L}{\partial \theta^s} = \frac{(-1)^{s-1}}{2\theta^{s+1}} n\{s\theta^2 - (s-1)\theta\} - \frac{n}{2}\delta_{s,1},$$

where the expectation is over the joint density of the sample, and δ is the Kronecker delta function. Note that the expectation of the first derivative is zero, and justify.

Example 1.5 The density of a random variate is
$$f(x, \theta) = 1 + \frac{(2x-1)}{6}\cos\theta + \frac{(3x-1)}{6}\sin\theta$$
$$(0 \leq x \leq 1, \quad 0 \leq \theta \leq 100\pi).$$

Investigate whether there is a consistent mle for θ in samples of n. If one such exists, $0 < \theta_0 < 2\pi$, show that there are 49 other solutions, each of which maximizes the likelihood. (This example is attributed to C.A.B. Smith; see Huzurbazar (1948).)

1.2.3 Bias

If, the expectation being over the joint density of the sample,

$$\text{E }\hat{\theta}_n(x_1, x_2, \ldots, x_n) = \theta, \quad \theta \in \Theta,$$

then $\hat{\theta}$ is unbiased for θ.

Unbiasedness is a desirable property, but nonetheless a very stringent restriction; for example, it is not even permitted for $\text{E}(\hat{\theta} - \theta) = A \exp(-n^\lambda)$, $\lambda > 0$, so that extensive classes of "almost unbiased" estimators are disqualified.

Since the early development of theoretical statistics concentrated mainly on properties of the binomial, Poisson, and normal distributions (even here m_2 is biased for σ^2), it is not surprising that modern estimation situations, where bias may be important, are not scrutinized carefully for this property. An argument against the study of bias seems to suggest that bias never was important, so that it can be ignored henceforth. A rejoinder to this is that if $\text{E }\hat{\theta} = \theta + \theta_1/n + \theta_2/n^2 + \ldots$, then how can we ignore $\theta_1, \theta_2, \ldots$ unless we assume (falsely) that they are zero or negligible?

However, there are devices which can be employed to reduce bias, one being due to Quenouille (Kendall and Stuart, 2, 5-8). Suppose, formally,

$$\text{E}(t_n) = \theta + \frac{a_1(\theta)}{n} + \frac{a_2(\theta)}{n^2} + \ldots.$$

Then, for a subset of $(n-1)$ observations,

$$\text{E}(t'_{n-1}) = \theta + \frac{a_1(\theta)}{n-1} + \frac{a_2(\theta)}{(n-1)^2} + \ldots$$

so that, eliminating the $a_1(\theta)$ term,

$$\text{E}(nt_n - \overline{n-1}\, t'_{n-1}) = \theta - \frac{a_2(\theta)}{n(n-1)} + \ldots$$

which is now biased to the second order in n^{-1}; clearly the same bias holds for the mean value of t'_{n-1} over all subsets of $(n-1)$.

Note in passing that if $\lim \text{E}(\hat{\theta}_n) = \theta$ as $n \to \infty$, then $\hat{\theta}_n$ is asymptotically unbiased for θ, provided θ is the mean of the asymptotic distribution.

Devices for reducing the bias and the effect of higher-order terms in variances, etc. are sometimes referred to as "jackknife" methods, the descriptive being introduced by Tukey (1958). Brillinger (1964) has discussed the application of the technique to the mean and variance of a maximum likelihood estimator. Gray and Schucany (1972) have discussed the generalized "jackknife" statistic.

Example 1.6 Let X have a Poisson distribution with mean θ, $\theta > 0$, so that

$$P_r(X = x) = e^{-\theta} \theta^x/x!, \quad x = 0, 1, \ldots,$$
$$= 0, \quad \text{otherwise.}$$

Show that the sample mean $\bar{x}$ is unbiased for θ. Is $\sqrt{\bar{x}}$ unbiased for $\sqrt{\theta}$? Anscombe (1948), in connection with variance stabilization, shows that

$$E\sqrt{(X + \tfrac{3}{8})} \sim \sqrt{(\theta + \tfrac{3}{8})} - \frac{1}{8\theta^{1/2}} + \frac{1}{64\theta^{3/2}} \quad (\theta \gg 0)$$

$$\text{Var}\sqrt{(X + \tfrac{3}{8})} \sim \frac{1}{4} + \frac{1}{64\theta^2}.$$

(Bartlett (1936) had previously shown that for large θ, $\sigma^2(\sqrt{X}) = \tfrac{1}{4}$.)

1.2.4 Asymptotic normality

A consistent solution of the likelihood equation

$$\frac{\partial \ln L(x|\theta)}{\partial \theta} = 0$$

is asymptotically normally distributed, with mean $v_1(\hat{\theta}) = \theta$ *(the true value) and variance*

$$v_2(\hat{\theta}) = \frac{1}{I(\theta)}, \tag{1.3}$$

where

$$I(\theta) = E\left\{-\frac{\partial^2 \ln L}{\partial \theta^2}\right\},$$

provided certain regularity conditions are satisfied.

There are so many ramifications of this result that it is almost impossible to consider many of the details; the reader should consult Kendall and Stuart, 2, 44–9; Cramér (1946), pp. 500–6; Rao (1952), pp. 157–61; Norden (1972).

The regularity conditions concern the existence and continuity of the first two derivatives of $\ln(L)$, their boundedness, and integrability of the third derivative. There is also an assumption that the probability density (or probability function) is such that

$$E\left\{\frac{\partial \log L}{\partial \theta}\right\}^2 = E\left\{-\frac{\partial^2 \log L}{\partial \theta^2}\right\},$$

which will be true if the density of the variate considered has a range of integration independent of θ.

It is also of interest to note three further theorems quoted by Rao, assuming the regularity conditions satisfied:

(i) The likelihood equation has, in probability, a solution which converges to the true value. (Note that this is a limiting property and may break down in small samples.)

(ii) A consistent solution of the likelihood equation provides (in probability) a maximum.

(iii) The variance $\nu_2(\theta)$ is the minimum possible asymptotic variance.

It is certainly of interest to know that mle are asymptotically normal (multivariate when simultaneous estimation is involved), but as is well known, the central limit theorem, in one or other of its forms, ensures the same for most estimators. For example, from Cramér (1946, pp. 353–66) *the random variable $f(m_r, m_s)$ is asymptotically normal with mean $f(\mu_r, \mu_s)$ and variance*

$$\left(\overline{\frac{\partial f}{\partial m_r}}\right)^2 \mu_2(m_r) + 2\, \overline{\frac{\partial f}{\partial m_r}}\, \overline{\frac{\partial f}{\partial m_s}}\, \mu_{11}(m_r, m_s) + \left(\overline{\frac{\partial f}{\partial m_s}}\right)^2 \mu_2(m_s),$$

where (i) m_r, m_s are sample central moments,

(ii) $\overline{\dfrac{\partial f}{\partial m_t}} \equiv \left.\dfrac{\partial f}{\partial m_t}\right|_{m_t = \mu_t}$,

(iii) $\mu_2(m_r) \equiv$ Variance (m_r) to order n^{-1}, and similarly for μ_{11},

provided that in a neighbourhood of $(m_r = \mu_r, m_s = \mu_s)$ the function $f(\cdot, \cdot)$ is continuous and has continuous derivatives of the first and second order with respect to m_r, m_s. Cramér goes on to remind us that functions such as $1/m_2$ are asymptotically normally distributed if the first four population moments exist, although the moments of $1/m_2$ in infinite samples may not exist.

1.2.5 MLE as efficient estimators

The fact that we are dealing with estimators which are asymptotically normal carries with it the point that the normal is completely determined by its mean (μ) and variance (σ^2); in particular, a deviation greater than $k\sigma$ from the mean has a probability of occurrence diminishing with k. Thus an estimator with smallest (or smaller) variance is to be preferred against others (it will be distributed closer to the true value than others); an estimator, if one such exists, with smallest variance is called an "efficient estimator".

The introduction of the concept of efficiency by Fisher (see for example his *Statistical Methods for Research Workers*, 1st edn, 1925) was intended as a large-sample property. The asymptotic variance (that is, the n^{-1} term) of the mle in general turns out to be less than or equal to the asymptotic variance of any competitor.

How small can the variance of estimators, belonging to the class of asymptotically unbiased estimators, become? The answer is contained in the so-called Cramér–Rao lower bound; this bound is exact and holds for finite samples.

If $E\,\hat{\theta}_n(x_1, x_2, \ldots, x_n) = \zeta(\theta)$, *then the variance* $V(\hat{\theta})$ *is not less than* $(d\zeta/d\theta)^2/I(\theta)$. The proof is a natural consequence of the Schwarz inequality for integrals; it follows from

$$\int \hat{\theta}_n(\mathbf{x}) f(\mathbf{x})\, d\mathbf{x} = \zeta(\theta),$$

and if differentiation under the integral sign is valid,

$$\int \hat{\theta}_n(\mathbf{x}) \frac{\partial f(\mathbf{x})}{\partial \theta}\, d\mathbf{x} = \zeta^{(1)}(\theta)$$

along with

$$\int (\hat{\theta}_n - \zeta)^2 f(\mathbf{x})\, d\mathbf{x} = V(\hat{\theta}),$$

where $f(\mathbf{x})$ refers to the joint density of the sample. When $\hat{\theta}_n$ is unbiased for θ, the bound reduces to $1/I(\theta)$.

One of us pointed out (Shenton, 1970) that this bound was very likely known to Aitken and Silverstone (1942); Freeman (1963, p. 243) remarks that it could be credited to Fisher, Fréchet, Darmois, and Dugué.

We digress momentarily to recall the main thrust of Aitken and Silverstone's idea (*for the single-parameter case*). Given that $(x_1, x_2, \ldots, x_n)$ are independent identically distributed random variates, each with density $\phi(\mathbf{x}; \theta)$, can a function of the sample $t_n(\mathbf{x})$ be found which is unbiased for θ and also has minimum variance over all samples? If this problem has simple solutions then a giant step forward would result. If $\phi(\mathbf{x}; \theta)$ is the joint density of the sample, then the aim is to solve simultaneously, for t_n and ϕ,

(i) $\quad \int \Phi(\mathbf{x}; \theta)\, d\mathbf{x} = 1$

(ii) $\quad \int t_n(\mathbf{x}; \theta)\, \Phi(\mathbf{x}; \theta)\, d\mathbf{x} = \theta$

(iii) $\quad \int (t_n(\mathbf{x}; \theta) - \theta)^2 \Phi(\mathbf{x}; \theta)\, d\mathbf{x} = \text{minimum}.$

If a solution exists, it will belong to the MVU class, minimum variance and unbiased. Aitken and Silverstone invoke concepts from the calculus of variations to show that there is a solution, provided that

$$\frac{\partial}{\partial \theta} \ln \Phi(\mathbf{x}; \theta) = (t_n(\mathbf{x}) - \theta)/\lambda(\theta),$$

where λ is independent of $\mathbf{x}$. One readily finds, for example, that $t = \Sigma\, x^2/n$ is a minimum-variance unbiased estimator for σ^2 in sampling from $N(0, \sigma^2)$; however, there is no such estimator for σ, so that this is evidence of the theorem's sensitivity to nonlinear mappings. Aitken and Silverstone go on to show that the variance V of the estimator is such that V^{-1} is the mean value of $(-\partial^2 \ln \Phi/\partial \theta^2)$.

In passing, it is of interest to note, as with many mathematical structures, that the MVU problem is much more difficult to solve for the multiparameter case (for example, in the remaining chapters it will be evident that second-order biases and covariances for ml estimators in the multiparameter case are dramatically more complicated than in the single-parameter situation, so much so that one's initial reaction leans towards the possibility of error in the heavy manipulative algebra involved, or the possibility of inherent redundancy). In a personal communication to one of us (7th March, 1959), Aitken remarked, in connection with the problem of estimating many parameters by unbiased minimum variance (Wilks's "generalized variance"): "... I am bound to say also that I found, for once, my own matrix theory almost too complicated for myself, especially round about pp. 373–375; and I think this has kept most people from reading the paper at all. I had had the whole idea some 12–15 years earlier, but was content in 1937 to let a pupil, xxxxx, work out the univariate consequences ...". Aitken was referring to his paper in *Proc. Roy. Soc. Edinb.*, **57**, 369–77.

1.2.6 Multiparameter estimation

Simultaneous estimation of parameters is commonplace, so some attention is now given to the required modifications in the results already quoted for the single-parameter case. Replacing θ by the vector $\boldsymbol{\theta} \equiv (\theta_1, \theta_2, \ldots, \theta_h)$, the search for solutions to the likelihood equation (1.2) in theory presents nothing new; however, in practice, iterative schemes for solutions may fail to converge (as may also be the case in single-parameter estimation), and the problem is greatly magnified when three or more parameters are estimated. In reality, the solution search belongs to techniques of numerical analysis, and these now co-exist with scientific computer programming. One need hardly remark that problems of solution searches are exacerbated as sample sizes become small (and here we must avoid being drawn into specifics about "smallness"). Kendall and Stuart (vol. 2, 54) remark that one searches for local turning points in

$$\frac{\partial \ln L(x, \boldsymbol{\theta})}{\partial \theta_j} = 0, \quad j = 1, 2, \ldots, h,$$

and with regularity conditions similar to those for the one-parameter case, they note that a sufficient condition for a turning point to be also a maximum is that the matrix $(\partial^2 \ln L/\partial \theta_i \partial \theta_j)$ be negative definite.

We next remark on the asymptotic normality in the many-parameter case. The covariance matrix, as in the transition from univariate to multivariate normal, now plays a role. The result is that under certain regularity conditions, the joint distribution of $(\hat{\theta}_1, \hat{\theta}_2, \ldots, \hat{\theta}_h)$ is asymptotically normal with covariance matrix

$$(v_{rs})^{-1} = \left(-E\frac{\partial^2 \ln L}{\partial \theta_i \partial \theta_j}\right)$$

$$= \left(E\frac{\partial \ln L}{\partial \theta_i}\frac{\partial \ln L}{\partial \theta_j}\right) \quad (i, j = 1, \ldots, h)$$

and mean $(\theta_1, \theta_2, \ldots, \theta_h)$. This result, from a formal Taylor-series point of view, is derived in Chapter 3. Note that the determinant of the covariance matrix defines generalized variance.

An interesting property is due to Geary (1942) but was very likely known to Fisher (1941). According to Geary:

In an infinite universe there are k frequency grades, k being arbitrary, the probability for the (i)th grade being p_i. The p_i are functions of known form of h parameters $\theta_1, \theta_2, \ldots, \theta_h$, of which the first partial derivatives all exist. In indefinitely large random samples the estimators of the parameter derived from the maximum likelihood equations are those which minimize the generalized variance.

The definition of asymptotic efficiency for simultaneous estimation follows from this result. For example, suppose the non-central sample moments m'_r, $r = 1, 2, \ldots, s$ estimate parameters θ_r, $r = 1, 2, \ldots, s$. Then the joint asymptotic efficiency (Fisher, 1941) E_f is given by

$$E_f^{-1} = \left|E\left(\frac{1}{f(x)}\frac{\partial f}{\partial \theta_j}\frac{\partial f}{\partial \theta_k}\right)\right| * |\text{Cov}(\theta_j^*, \theta_k^*)|$$

$$\sim \{J(\theta/\mu')\}^2 \left|E\left(\frac{1}{f(x)}\frac{\partial f}{\partial \theta_j}\frac{\partial f}{\partial \theta_k}\right)\right| * |\text{Cov}(m'_j, m'_k)|,$$

where θ_j^* are consistent for θ_j, and J is the Jacobian

$$J(\theta/\mu') = \frac{\partial(\theta_1, \theta_2, \ldots, \theta_s)}{\partial(\mu'_1, \mu'_2, \ldots, \mu'_s)}.$$

For example, Katti and Gurland (1962) give the asymptotic efficiency of the method of moments against ml, in estimating the parameters p, k of the negative binomial distribution (pgf; $(1 + p - pt)^{-k}$, $p, k > 0$). A few illustrations are given in Table 1.1.

Table 1.1
Asymptotic efficiency for the negative binomial (moments v. ml)

k \ p	0·2	1·0	10·0
0·5	0·909	0·713	0·399
2·0	0·943	0·821	0·635
10·0	0·981	0·942	0·892

Is it really true that there is little lost in using the much simpler moment estimators when p is small and k large, or have we overlooked something? Actually we have ignored the fact that the results are purely asymptotic and are only valid in so far as first-order asymptotic covariances are good approximations to true covariances — in other words, the results are dangerous to accept unless there is a restriction on the sample size.

Example 1.7 Estimators t_1 and t_2 are functions of the sample moments m_1, m_2, and

$$t_1 = f_1(m_1, m_2), \quad t_2 = f_2(m_1, m_2).$$

Working to first-order asymptotics, find an expression for

$$\begin{vmatrix} \text{Var } t_1 & \text{Cov }(t_1, t_2) \\ \text{Cov }(t_2, t_1) & \text{Var } t_2 \end{vmatrix}$$

in terms of the first three population moments μ_2, μ_3, μ_4.

For the estimation of the parameters ρ, a from a random sample of n from the gamma density,

$$f(x; a, \rho) = e^{-x/a} (x/a)^{\rho-1}/a\, \Gamma(\rho), \quad x > 0,$$
$$= 0, \quad x < 0,$$

where $a, \rho > 0$, moment estimators a^*, ρ^* are used, where

$$a^* = m_2/m_1, \quad \rho^* = m_1^2/m_2.$$

Using the results (Kendall and Stuart, vol. 1, **10.5**)

$$\text{Var } m_1 = \mu_2/n, \quad \text{Cov }(m_1, m_2) \sim \mu_3/n, \quad \text{Var } m_2 \sim (\mu_4 - \mu_2^2)/n,$$

evaluate the dominant asymptotic term in

$$\begin{vmatrix} \text{Var } a^* & \text{Cov }(a^*, \rho^*) \\ \text{Cov }(\rho^*, a^*) & \text{Var } \rho^* \end{vmatrix}$$

Using the asymptotic covariance matrix for mle, evaluate the dominant term in the asymptotic joint efficiency, showing that

$$E_f = \frac{1}{2(\rho+1)(\rho\psi^{(1)}(\rho) - 1)},$$

where $\psi^{(1)}(\rho) = d^2 \ln \Gamma(\rho)/d\rho^2$ is the derivative of the psi function. Show that for large ρ, the asymptotic efficiency is

$$E_f \sim 1 - \frac{13}{\rho}.$$

1.2.7 Sufficiency

Let $\theta, \theta_1^*, \theta_2^*, \ldots, \theta_r^*$, $r = 1, 2, \ldots, n-1$, $(n \geq 2)$, be a set of functionally independent statistics, each an estimator of a parameter θ. For any choice of θ_j^* ($j = 1$ to $n - 1$) whatever, let the joint density of these estimators factor into two parts, i.e.

$$f(\hat{\theta}, \theta_1^*, \theta_2^*, \ldots, \theta_r^* | \theta) = f_1(\hat{\theta} | \theta) f_2(\theta_1^*, \theta_2^*, \ldots, \theta_r^* | \hat{\theta}),$$

where the second factor is a conditional density independent of θ. Then $\hat{\theta}$ is said to be sufficient for θ, for the distribution of $\theta_1^*, \theta_2^*, \ldots$ adds no further information to that supplied by $\hat{\theta}$.

For example (Kendall and Stuart, vol. 2, 27–8), if ϕ is known, then $\hat{\theta} = \sqrt{\{\Sigma (x - \phi)^2/n\}}$ is sufficient for σ in sampling from $N(\phi, \sigma^2)$.

Joint sufficiency in the case of a multivariate distribution is a natural extension of the definition in the univariate single-parameter case.

If, for any choice of $\boldsymbol{\theta}^*$ whatever, the joint density of a set of functionally independent vector estimators factors into a part $f_1(\hat{\boldsymbol{\theta}}|\boldsymbol{\theta})$ and a conditional density $f_2(\boldsymbol{\theta}^*|\hat{\boldsymbol{\theta}})$, then $\hat{\boldsymbol{\theta}}$ is said to be sufficient for $\boldsymbol{\theta}$.

For the two-parameter gamma density (considered in detail in Chapter 5) with scale parameter a, shape parameter ρ, the mle's are jointly sufficient for a and ρ.

Again, in estimating the parameter θ for the density $f(x; \theta)$ (where $\theta \in \Theta$), it can be shown that if $\tilde{\theta}(x_1, x_2, \ldots, x_n)$ is a sufficient statistic for θ, and if the mle $\hat{\theta}$ of θ is unique, then $\hat{\theta}$ is a function of $\tilde{\theta}$.

Much confusion has arisen in the literature because of different schools of thought pressing claims for *best estimators* through an over-enthusiastic interpretation of such properties as sufficiency. E.S. Pearson (1974) has pointed out that sufficiency can only be an advantageous property if the hypothesized distributional form is true.

Sufficient statistics with only one or two components do not usually seem to exist in practical situations; this is in no way intended to detract from the theoretical attractiveness of the concept, but to emphasize that the property is restrictive and adds little to our knowledge of the distributional properties of the estimators.

1.3 Small-sample problems

1.3.1 Introduction

The standard theory of mle we have outlined needs careful scrutiny when applied to small samples, for much of the theory is basically asymptotic. We need to know how misleading the formula is for the asymptotic variance $v_2(\hat{\theta}) = 1/I(\theta)$ (see *1.2.4*) when the sample size is, say, 50 or 100; similarly, how serious is the bias, and how important are the departures from asymptotic normality when questions relating to errors of estimates and inferences on parameters are answered.

To understand fully the remaining chapters, some clear ideas of what is meant by asymptotic results are essential. Although many scientists frequently use asymptotic methods, the subject is still rather ill-defined and difficult to generalize. Again, a glance at the general formulae for the n^{-2} terms in the bias and variance of ml estimators given in Chapter 2 (and also the corresponding formulae in the multiparameter situation in Chapter 3) is almost certain to lead to the question "why do they look so complicated?". Is there not a more direct approach?

Lastly, the fact is that ml estimators have few properties which hold for all sample sizes — nonetheless the method of mle has wide applicability and general usefulness, so that a closer look at its behaviour in the practical world of small samples is essential.

1.3.2 Three illustrations

(a) *The Poisson distribution*   For this case, with

$$\Pr(X = x) = e^{-\theta} \theta^x / x! \quad x = 0, 1, \ldots,$$
$$(\theta > 0)$$

the random sample resulting in n_j observations of $x = j$ $(j = 0, 1, \ldots)$, the likelihood function is proportional to

$$L = e^{-n\theta} \theta^{n\bar{x}}$$

with
$$\begin{cases} \dfrac{\partial \log L}{\partial \theta} = -n + \dfrac{n\bar{x}}{\theta}, \\[2mm] -\dfrac{\partial^2 \log L}{\partial \theta^2} = \dfrac{n\bar{x}}{\theta^2}. \end{cases}$$

Thus $\hat{\theta} = \bar{x}$, and from (1.3)

$$\operatorname{Var} \hat{\theta} \sim \theta/n,$$

and this first-order asymptotic happens to be exact.

Notice, notwithstanding the well-known properties of sufficiency, etc. in this case, that $\partial \log L/\partial \theta$ reduces to a simple statistic, so that the equation for $\hat{\theta}$ is explicit and in one dimension (that of $\bar{x}$). Moreover the distribution of $\hat{\theta}$ is that of the mean, and is precisely known in this case, being a Poisson variate with mean θ and taking the values (x/n), $x = 0, 1, \ldots$.

(b)   *Zero-truncated Poisson*   Here

$$\Pr(X = x) = \frac{e^{-\theta} \theta^x}{x!(1 - e^{-\theta})}, \quad \begin{array}{l} x = 1, 2, \ldots \\ (\theta > 0) \end{array} \tag{1.4}$$

and for a sample of n,

$$\begin{cases} \dfrac{\partial \log L}{\partial \theta} = -n + \dfrac{n\bar{x}}{\theta} - \dfrac{n}{e^\theta - 1}, \\ -\dfrac{\partial^2 \log L}{\partial \theta^2} = \dfrac{n\bar{x}}{\theta^2} - \dfrac{ne^\theta}{(e^\theta - 1)^2}. \end{cases} \quad (1.5)$$

Thus

$$\hat{\theta}/(1 - e^{-\hat{\theta}}) = \bar{x}, \quad (1.6)$$

which clearly leads to $\hat{\theta}$ uniquely. Since the mean of the distribution is $\mu_1' = \theta/(1 - e^{-\theta})$, the mle is the same as the simplest moment estimation. Notice that the asymptotic variance is

$$\operatorname{Var} \hat{\theta} \sim \frac{\theta(1 - e^{-\theta})^2}{n(1 - e^{-\theta} - \theta e^{-\theta})} \quad (1.7)$$

found from (1.3) or from the usual incremental approach, namely,

$$\left\{ \frac{1}{1 - e^{-\theta}} - \frac{\theta e^{-\theta}}{(1 - e^{-\theta})^2} \right\} \delta\hat{\theta} \sim \delta\bar{x}$$

so

$$E(\delta\hat{\theta})^2 \sim \frac{\operatorname{Var} \bar{x}}{(1 - e^{-\theta} - \theta e^{-\theta})/(1 - e^{-\theta})^2},$$

as before. More information on $\hat{\theta}$ arises from inversion of the equation defining $\hat{\theta}$ or from the likelihood function; these lead to the same series for $\hat{\theta}$ in terms of powers of $\bar{x} - \mu_1'$. Thus, formally,

$$\hat{\theta} = \sum_{s=0}^{\infty} A_s (\bar{x} - \mu_1')^s / s! \quad (1.8)$$

Now if in (1.6) $\bar{x} \to \mu_1'$, then $\hat{\theta} \to \theta$. Hence

$$A_s = \left. \frac{\partial^s \hat{\theta}}{\partial \bar{x}^s} \right|_{\bar{x} = \mu_1', \, \hat{\theta} = \theta}$$

giving $A_0 = \theta$,

$$A_1 = \left\{ \frac{1 - e^{-\theta} - \theta e^{-\theta}}{(1 - e^{-\theta})^2} \right\}^{-1},$$

$$A_2 = \left\{ \frac{2e^{-\theta} - \theta e^{-\theta}}{(1 - e^{-\theta})^2} - \frac{2\theta\, e^{-2\theta}}{(1 - e^{-\theta})^3} \right\} A_1^3,$$

and so on, the coefficients becoming more complicated and being found by repeated differentiation of a function of $\hat{\theta}$ (Faà di Bruno's formula; see *5.2.4* in Chapter 5). Now noting that the sampling moments of the mean are, for example,

$$\mu'_1(\bar{x}) = \mu'_1,$$
$$\mu_2(\bar{x}) = \mu_2/n,$$
$$\mu_3(\bar{x}) = \mu_3/n^2,$$
$$\mu_4(\bar{x}) = 3\mu_2^2/n^2 + (\mu_4 - 3\mu_2^2)/n^3,$$
$$\mu_5(\bar{x}) = 10\mu_2\mu_3/n^3 + (\mu_5 - 10\mu_2\mu_3)/n^4,$$
$$\mu_6(\bar{x}) = 15\mu_2^3/n^3 + (10\mu_3^2 + 15\mu_4\mu_2 - 45\mu_2^3)/n^4$$
$$+ (\mu_6 - 15\mu_4\mu_2 - 10\mu_3^2 + 30\mu_2^3)/n^5,$$

we find from (1.8)

$$E(\hat{\theta}) = \theta + \frac{A_2\mu_2}{2!\,n} + \frac{1}{n^2}\left(\frac{A_3\mu_3}{3!} + \frac{3A_4\mu_2^2}{4!}\right)$$
$$+ \frac{1}{n^3}\left(\frac{A_4(\mu_4 - 3\mu_2^2)}{4!} + \frac{10A_5\mu_2\mu_3}{5!} + \frac{15A_6\mu_2^3}{6!}\right) + \ldots . \quad (1.9)$$

Similarly for the variance, using

$$\text{Var}\,\hat{\theta} = E(\hat{\theta} - \theta)^2 - (E(\hat{\theta} - \theta))^2,$$

we have from (1.8), formally,

$$E(\hat{\theta} - \theta)^2 = A_1^2\mu_2(\bar{x}) + A_1A_2\mu_3(\bar{x}) + \left(\frac{A_2^2}{4} + \frac{A_1A_3}{3}\right)\mu_4(\bar{x}) + \ldots ,$$

and assuming the validity of rearranging the terms,

$$E(\hat{\theta} - \theta)^2 = \frac{A_1^2\mu_2}{n} + \frac{1}{n^2}\left(A_1A_2\mu_3 + 3\left(\frac{A_2^2}{4} + \frac{A_1A_3}{3}\right)\mu_2^2\right) + \ldots ,$$

from which

$$\text{Var}\,\hat{\theta} = \frac{A_1^2\mu_2}{n} + \frac{1}{n^2}\left(A_1A_2\mu_3 + \left(A_1A_3 + \frac{A_2^2}{2}\right)\right) + \ldots . \quad (1.10)$$

Similarly, higher-order terms could be found; also the skewness $\sqrt{\beta_1}\,(= \mu_3(\hat{\theta})/\mu_2^{3/2}(\hat{\theta}))$ and the kurtosis $\beta_2\,(= \mu_4(\hat{\theta})/\mu_2^2(\hat{\theta}))$.

The expansions (1.9) and (1.10) are asymptotic in the sample size n in the sense that, for example,

$$\lim_{n\to\infty}\{E(\hat{\theta}) - \theta\} = 0,$$

$$\lim_{n\to\infty} n\left\{E(\hat{\theta}) - \theta - \frac{A_2\mu_2}{2n}\right\} = 0,$$

and in general, writing

$$E(\hat{\theta}) = \theta + \sum_{s=1}^{\infty} \zeta_s(\theta)/n^s, \tag{1.11}$$

where $\zeta(\theta)$ is a function of θ only,

$$\lim_{n \to \infty} n^s \left\{ E(\hat{\theta}) - \theta - \sum_{r=1}^{s+1} \zeta_r(\theta)/n^r \right\} = 0, \quad s = 1, 2, \ldots. \tag{1.12}$$

As we shall note in the sequel, a finite number of terms of series such as (1.11) may be used, for large enough n, to assess $E(\hat{\theta})$. The most favourable case for series of this kind arises when the error committed in using a finite set of terms is less in magnitude than the first term omitted.

The general formulae of Chapter 2 applied to this case yield higher-order terms for the first four moments (Tables 1.2 and 1.3).

Table 1.2
Bias and variance of θ

	θ	Coefficient of					
		n^{-1}	n^{-2}	n^{-3}	n^{-4}	n^{-5}	n^{-6}
Bias	0·1	−0·0312	0·0183	−0·0100	0·0031	0·0033	−0·0085
	1·0	−0·1726	0·0237	0·0247	−0·0047	−0·0262	−0·0046
	10·0	−0·0018	−0·0029	−0·0021	−0·0004	0·0002	−0·0002
Var $\hat{\theta}$	0·1	0·1936	−0·1108	0·0673	−0·0208	−0·0293	0·0728
	1·0	1·5122	−0·1217	−0·2054	0·0204	0·2428	0·0876
	10·0	10·0041	0·0282	0·0358	0·0157	−0·0030	0·0002

(Note: $E(\hat{\theta} - \theta)$ defines the bias here.)

Table 1.3
$\mu_3(\hat{\theta})$ and $\mu_4(\hat{\theta})$

	θ	Coefficient of				
		n^{-2}	n^{-3}	n^{-4}	n^{-5}	n^{-6}
$\mu_3(\hat{\theta})$	0·1	0·3505	−0·2403	0·0739	0·1607	−0·3880
	1·0	1·2428	0·9587	−0·0276	−1·3906	−0·8235
	10·0	9·9354	−0·2734	−0·2378	−0·0064	0·0396
$\mu_4(\hat{\theta})$	0·1	0·1124	0·3953	−0·0457	−0·6827	1·4976
	1·0	0·8599	−2·1227	−1·6649	5·9020	7·2614
	10·0	300·2453	12·3362	3·9103	1·5030	−0·8287

(The n^{-4} through n^{-6} terms were found following the approach of 5.3 in Chapter 5.)

The higher-order terms are evidently almost negligible for the particular parameter space studied, and this knowledge leads to increased confidence in the first-order asymptotics. The skewness and kurtosis (Table 1.4) were derived from moment expansions including terms of order n^{-6}. They are evidently markedly different from the first-order asymptotics results for θ and n small, but tend to merge for increasing θ and increasing n.

Table 1.4
Skewness and kurtosis

Sample size (n)	5		10		25		50	
θ	$\sqrt{\beta_1}$	β_2	$\sqrt{\beta_1}$	β_2	$\sqrt{\beta_1}$	β_2	$\sqrt{\beta_1}$	β_2
0·1	1·39	6·15	1·32	4·50	0·83	3·58	0·53	3·28
	(2·24)	(3·00)	(1·30)	(3·00)	(0·82)	(3·09)	(0·58)	(3·00)
1·0	0·35	2·93	0·22	2·96	0·14	2·98	0·10	2·99
	(0·30)	(3·00)	(0·21)	(3·00)	(0·13)	(3·00)	(0·09)	(3·00)
10·0	0·14	3·02	0·10	3·01	0·06	3·00	0·04	3·00
	(0·14)	(3·00)	(0·10)	(3·00)	(0·06)	(3·00)	(0·04)	(3·00)

(Parenthetic entries refer to first-order asymptotic values.)

The problem of evaluating the moments of $\hat{\theta}$ in this case is a step higher in difficulty than in case (a), since here, although the dimensionality is still one, the estimating equation (1.6) is implicit in $\hat{\theta}$. Note also that terms of order n^{-2} in $E(\hat{\theta})$ arise from $\mu_3(\bar{x})$, $\mu_4(\bar{x})$; terms of order n^{-3} arise from $\mu_4(\bar{x})$, $\mu_5(\bar{x})$, $\mu_6(\bar{x})$; and so on. Thus contributions to the expectation of a power of $\hat{\theta} - \theta$ arise, in general, from more than one term in the power series itself.

A more rigorous treatment of the problem would be to assess a bound for the expression in the braces in (1.12) for θ in some interval; however, even in this very simple case there are already serious difficulties.

(Note that the usual approach to the moments of $\hat{\theta}$, through the likelihood function, would proceed as follows. From (1.5),

$$\frac{\partial \log L}{\partial \hat{\theta}} = \frac{\partial \log L}{\partial \theta} + (\hat{\theta} - \theta)\frac{\partial^2 \log L}{\partial \theta^2} + \frac{(\hat{\theta} - \theta)^2}{2!}\frac{\partial^3 \log L}{\partial \theta^3} + \cdots,$$

where

$$\frac{\partial \log L}{\partial \theta} = -n + \frac{n\bar{x}}{\theta} - \frac{n}{e^\theta - 1},$$

$$\frac{\partial^2 \log L}{\partial \theta^2} = -\frac{n\bar{x}}{\theta^2} + \frac{ne^\theta}{(e^\theta - 1)^2},$$

$$\frac{\partial^3 \log L}{\partial \theta^3} = \frac{2n\bar{x}}{\theta^3} + n\left(\frac{e^\theta}{(e^\theta - 1)^2} - \frac{2e^{2\theta}}{(e^\theta - 1)^3}\right).$$

The point to note is that not only is the series for $\partial \log L/\partial \hat{\theta}$ to be inverted, but there is the additional problem that the coefficients in the inverted series involve the statistic $\bar{x}$ in some form in the denominators. This throws some light on the complicated nature of the moment expansions for $E(\hat{\theta})$ and $\text{Var}\,\hat{\theta}$ in the general case.)

(c) *The Cauchy density* Consider the estimation of θ for the density

$$f(x;\theta) = \frac{1}{\pi[1+(x-\theta)^2]}, \quad -\infty < x < \infty \quad (\theta \text{ real}).$$

Here the moments higher than the mean do not exist, and the mean itself only exists if defined as a principal value of the integral. In addition, it is well known that the sample mean $\bar{x}$ has exactly the same distribution as the original variate. If we are interested in a simple estimator to replace the mean, then the median of the sample would be suitable (the density of the median has been given by Fisher, 1925, p. 715).

The mle based on the sample $(x_1, x_2, \ldots, x_n)$ would be a solution of

$$\frac{\partial \log L}{\partial \theta} = 2 \sum \frac{(x_j - \theta)}{[1 + (x_j - \theta)^2]}$$

$$= 0.$$

If this were rationalized, the resulting equation in θ would be of degree $2n - 1$, so that a solution $\hat{\theta}$ would be a complicated function of $x_1, x_2, \ldots, x_n$ and of dimension n, in marked contrast to the one-dimensionality of cases (a) and (b). The asymptotic formula for Var $\hat{\theta}$ is now a powerful tool, and since

$$-\frac{\partial^2 \log L}{\partial \theta^2} = 2 \sum \left\{ \frac{1}{[1+(x_j-\theta)^2]} - \frac{2(x_j-\theta)^2}{[1+(x_j-\theta)^2]^2} \right\}$$

we see that asymptotically,

$$\text{Var } \hat{\theta} \sim \frac{2}{n}.$$

However, higher-order terms are found only after some non-trivial algebraic manipulations. Merely as an illustration, and to highlight the nature of the problem, consider the expansion for $\hat{\theta}$. We have

$$\frac{\partial \log L}{\partial \hat{\theta}} = 0$$

$$= \frac{\partial \log L}{\partial \theta} + (\hat{\theta} - \theta) \frac{\partial^2 \log L}{\partial \theta^2} + \frac{(\hat{\theta}-\theta)^2}{2!} \frac{\partial^3 \log L}{\partial \theta^3}$$

$$+ \frac{(\hat{\theta}-\theta)^3}{3!} \frac{\partial^4 \log L}{\partial \theta^4} + \ldots, \quad (1.13)$$

where

$$\frac{\partial \log L}{\partial \theta} = 2 \sum \frac{y}{D} \quad \begin{aligned} &(y \equiv y_j = x_j - \theta) \\ &(D \equiv D_j = 1 + y_j^2) \end{aligned}$$

$$\frac{\partial^2 \log L}{\partial \theta^2} = \sum \left\{ \frac{2}{D} - \frac{4}{D^2} \right\}$$

$$\frac{\partial^3 \log L}{\partial \theta^3} = \sum \left\{ \frac{4y}{D^2} - \frac{16y}{D^3} \right\}$$

$$\frac{\partial^4 \log L}{\partial \theta^4} = \sum \left\{ \frac{12}{D^2} - \frac{96}{D^3} + \frac{96}{D^4} \right\}.$$

Define $\partial^s \log L / \partial \hat{\theta} = L_s$, so that (1.13) can be formulated as

$$\hat{\theta} = \theta + \frac{(-L_1)}{L_2 + \frac{(\hat{\theta} - \theta)}{2} L_3 + \frac{(\hat{\theta} - \theta)^2}{6} L_4 + \ldots} \tag{1.14}$$

which now subscribes to the pattern of Lagrange's equation $\hat{\theta} = \theta + \chi(\hat{\theta})$ (Whittaker and Watson, 1915). Thus, formally

$$\hat{\theta} = \theta + \sum_{s=1}^{\infty} \left\{ \frac{1}{s!} \frac{d^{s-1}}{d\hat{\theta}^{s-1}} \chi^s(\hat{\theta}) \right\}_{\hat{\theta} = \theta}. \tag{1.15}$$

It is wise to relate the inversion of (1.13) to an established theorem, for structural errors in any *ad hoc* approach are a distinct possibility. A few terms of (1.15) are:

$$\hat{\theta} = \theta - \frac{L_1}{L_2} - \frac{L_1^2 L_3}{2! L_2^2} + \frac{L_1^3}{3! L_2^3} \left(\frac{L_4}{L_2} - \frac{3 L_3^2}{L_2^2} \right)$$

$$+ \frac{L_1^4}{4! L_2^4} \left(-\frac{L_5}{L_2} + \frac{10 L_3 L_4}{L_2^2} - \frac{15 L_3^3}{L_2^3} \right) + \ldots. \tag{1.16}$$

Notice that since $E(L_1) = 0$, L_1 acts as a pivotal element in the expansion and in a probability sense is the controlling term in the order of magnitude of successive terms after taking expectations.

There is still the necessity to rid the denominator terms (all of which are powers of L_n) of random elements, and this is done by setting

$$L_2 = \mathcal{L}_2 + \ell_2,$$

where $\mathcal{L}_2 = E(L_2)$ is in general non-zero and relates to the asymptotic variance. Finally one must satisfy oneself that a term such as

$$E(L_1^{s_1} L_2^{s_2} \ldots L_m^{s_m})$$

has lowest and highest terms in n^{-1} of orders $[\frac{1}{2} \Sigma s_\lambda]$ and $(-1 + \Sigma s_\lambda)$ respectively; the terms in (1.16) can now be arranged in descending powers of n^{-1}, and to include terms in $E(\hat{\theta})$ as far as n^{-2}, n^{-3}, etc. we must include the terms in (1.16) involving L_1^4, L_1^6, etc., respectively.

Example 1.8 The reader may return to this example after reading as far as paragraph 2.7.6 in Chapter 2. To assist in the evaluation of the various expectations involved, such as

$$[1^\alpha 2^\beta \ldots] = E\left\{\left(\frac{\partial \ln f}{\partial \theta}\right)^\alpha \left(\frac{\partial^2 \ln f}{\partial \theta^2}\right)^\beta \ldots\right\},$$

define $y = \theta - x$, $d = 1 + y^2$, and prove that if

$$\Gamma_s = \frac{\partial^s}{\partial \theta^s} \ln\left(\frac{1}{\pi d}\right),$$

then

$$\begin{cases} \Gamma_{2s} = \frac{2(2s-1)!}{d^{2s}} y_{2s}, & s = 1, 2, \ldots, \\ \Gamma_{2s+1} = \frac{2(2s)!}{d^{2s+1}} y\, y_{2s+1}, & s = 0, 1, \ldots, \end{cases}$$

where

$$\begin{cases} y_{2s+1} + 2y_{2s} + dy_{2s-1} = 0 & (s \geq 1) \\ y_{2s} + 2(d-1)y_{2s-1} + dy_{2s-2} = -d\delta_{s,1} \end{cases}$$

and $y_0 = 0$, $y_1 = -1$, δ the Kronecker function. For example,

$$\Gamma_1 = -2y/d, \quad \Gamma_2 = (2d-4)/d^2,$$
$$\Gamma_3 = (-4d+16)y/d^3, \quad \Gamma_4 = (12d^2 - 96d + 96)/d^4.$$

In addition, prove that

$$E\, d^{-s} = \frac{1.3 \ldots (2s-1)}{2^s s!} \quad (s = 1, 2, \ldots)$$

and deduce that

(i) $E(\Gamma_r \Gamma_s) = 0$ if $r + s$ is odd,

(ii) $E(\Gamma_2) = -\frac{1}{2}$, $E(\Gamma_2^2) = \frac{7}{8}$, $E(\Gamma_1 \Gamma_3) = -\frac{3}{4}$,
$E(\Gamma_1^2 \Gamma_2) = -\frac{1}{8}$, $E(\Gamma_4) = \frac{3}{4}$.

From Chapter 2 (or otherwise) show that the n^{-1}, n^{-2} terms in $E(\hat{\theta})$ are zero, and that

$$\text{Var}\,\hat{\theta} \sim \frac{2}{n} + \frac{5}{n^2}.$$

As a comparison with the mle, consider the median estimator $\tilde{\theta}$. The density is known (Fisher, 1925) to be

$$M(\tilde{x}; \theta) = \frac{(2m+1)!}{(m!)^2 \pi^{2m+1}} \left(\frac{\pi^2}{4} - \phi^2\right)^m \frac{1}{1 + (\tilde{x} - \theta)^2}$$

for a sample of $2m + 1$, where $\tan \phi = \tilde{x} - \theta$, $|\phi| < \tfrac{1}{2}\pi$. Deduce

(i) $E(\tilde{x}) = \theta$

(ii) $\operatorname{Var} \tilde{x} = \dfrac{(2m+1)!}{(m!)^2 \, 2^{2m}} \int_0^1 (1-x^2)^m \tan^2(\tfrac{1}{2}\pi x)\, dx$.

Using the series
$$\tan y = \sum_{s=1}^{\infty} b_s y^{2s+1} \quad (y^2 < \pi^2/4),$$

where
$$b_s = \dfrac{2^{2s}(2^{2s}-1)}{(2s)!} |B_{2s}|,$$

and B_s is a Bernoulli number ($B_0 = 1$, $B_1 = -\tfrac{1}{2}$, $B_2 = \tfrac{1}{6}$, $B_4 = -\tfrac{1}{30}$, etc.), show that

$$\operatorname{Var} \tilde{x} \sim \sum_{s=1}^{\infty} \dfrac{[1 \cdot 3 \cdot \ldots \cdot (2s-1)](\pi/2)^{2s} d_s}{[(2m+3)(2m+5) \ldots (2m+2s+1)]},$$

where
$$d_s = b_{(s+1)/2}^2 + 2 \sum_{r=1}^{[s/2]} b_r b_{s+1-r},$$

and $b_r = 0$ if r is not an integer, $[x]$ being the integer part of x. In particular,

$$\operatorname{Var} \tilde{x} \sim \dfrac{\pi^2}{4(2m+3)} + \dfrac{\pi^4}{8(2m+3)(2m+5)} + \dfrac{17\pi^6}{192(2m+3)(2m+5)(2m+7)}.$$

Numerically the series is asymptotically benign; thus for $m = 10$,

$$\operatorname{Var} \tilde{x} \sim 0{\cdot}107278 + 0{\cdot}021176 + 0{\cdot}005483 + 0{\cdot}001701$$
$$+ 0{\cdot}000604 + 0{\cdot}000230$$
$$= 0{\cdot}136480.$$

The so-called asymptotic efficiency is clearly $2/(\pi^2/4) = 8/\pi^2 = 0{\cdot}81$. Inclusion of second-order terms leads to the efficiency $(8/\pi^2)(1 - 2{\cdot}4/n)$.

Example 1.9 Derive the terms in (1.16) by assuming
$$\hat{\theta} = \sum_{s=0}^{\infty} a_s L_1^s$$

and evaluating $\partial^r \hat{\theta}/\partial L_1^r$ from (1.13).

Also derive the terms by assuming
$$\hat{\theta} = A_1 + A_2 + A_3 + \ldots$$

and substituting in (1.13), and equating to zero terms of the same order of magnitude, the assumption being that A_s is a term of magnitude s in terms of L_1, the lowest order.

Example 1.10 In estimating two parameters θ_1, θ_2 by ml, show that formal solutions to

$$\begin{cases} L_{10}(\hat{\theta}_1, \hat{\theta}_2) = \dfrac{\partial \log L}{\partial \hat{\theta}_1} = 0 \\[6pt] L_{01}(\hat{\theta}_1, \hat{\theta}_2) = \dfrac{\partial \log L}{\partial \hat{\theta}_2} = 0 \end{cases}$$

may be derived by assuming

$$\begin{cases} \hat{\theta}_1 - \theta_1 = A_1 + A_2 + \ldots \\ \hat{\theta}_2 - \theta_2 = B_1 + B_2 + \ldots \end{cases}$$

(where A_1, B_1 are first order, A_2, B_2 second order, etc.) and substitution in $L_{10} = L_{01} = 0$.

Find expressions for A_s, B_s, $s = 1, 2, 3, 4$ and indicate how expectations of these terms could be evaluated.

1.3.3 The above illustrations ((a), (b) and (c)) highlight the problems involved in evaluating higher-order terms in the moments of a single mle, and give some idea of what may be involved in the multiparameter case.

The reader will see that the difficulties are of two kinds. In the first place, there is the problem of the dimensionality of the Taylor series for $\hat{\theta}$, and this may be any number not exceeding the sample size. Complications are alleviated in this respect if the equation for $\hat{\theta}$ is explicit. The second kind of difficulty arises at the expectation stage, for each term in the Taylor expansion involves random variables in its denominator, and also, in general, each term generates more than one contribution to a coefficient of a power of n^{-1} in an expectation such as $E(\hat{\theta} - \theta)$.

In Chapter 2 we derive expansions for the first four moments of $\hat{\theta}$ in the general case, i.e. no simplifying assumptions as to dimensionality (or implicitness) are made. The derivation, for simplicity, is embedded in a discrete probability structure, but the final expressions are in terms of expectations of products of derivatives of the probability function (or density).

Chapter 3 gives expressions for the n^{-2} biases and covariances in the multiparameter case; again, the final expressions involve similar terms to the single-parameter case.

Special density functions (such as *1.3.2* (b)) are treated in Chapter 5.

1.3.4 *Summary of types of estimation problems*

There are at least three classes of equations which arise:

(a) *Estimators given explicitly in terms of simple functions of the sample values*

Examples: As already mentioned in *1.3.2*(a), the sample mean is the ml estimator of θ for the Poisson distribution with pgf $\exp\theta(t-1)$. The standard deviation in sampling from $N(0, \sigma^2)$ is estimated by $\sqrt{m_2}$; odd moments of this are infinite series if we use the Taylor series approach, although exact moments in terms of gamma functions are readily derived.

(b) *Estimators given implicitly in terms of a fixed number of statistics, independent of the sample size*

Examples: (i) the scale and shape parameters of the two-parameter gamma density are implicit functions of the ratio (Sample mean/Sample geometric mean), (see Chapter 5). An additional, highly desirable property holds here, for the estimator of the scale parameter divided by the scale parameter is distributed independently of the scale parameter (other examples are given by Antle and Bain (1969)). (ii) parameters of the logarithmic series distribution, zero-truncated binomial distribution, zero-truncated Poisson distribution (see *1.3.2*(b)), are estimated by implicit functions of the sample mean.

(c) *Estimators given implicitly in the form $f(\theta; x_1, x_2, \ldots, x_N) = 0$ in the single-parameter case, and $f_i(\theta; x) = 0$, $i = 1, 2, \ldots, h$ for h parameters*

Examples: (i) $\hat{\theta}$ in the Cauchy density is implicit and of dimension n (see *1.3.2*(c)). (ii) the parameters s, a, ρ for the three-parameter gamma density, where s is a terminal of the distribution, a and ρ referring to scale and shape parameters, are estimated implicitly in terms of an infinite number of functions of the sample values. (iii) many cases occurring in practice are of this type; examples are the negative binomial, Pólya–Aeppli, Neyman Type A, Weibull, Pearson Types I and IV, and many others.

In general, it is impossible to write down the general term (in descending powers of N) in a moment of an estimator, although it may be defined by a finite sequence of inter-related operations. For estimation structures of class (c), we can at present evaluate N^{-1} and N^{-2} biases and covariances; N^{-2} terms provide invaluable information on the type of series involved and may warn of divergent tendencies.

Since so-called divergent series play a major role in the development of the subject, we next review briefly the relevant part of the theory.

1.3.5 Asymptotic series

Define
$$R_r(x) = f(x) - \phi(x) \sum_{s=0}^{r} \frac{A_s}{x^s}.$$

Then $f(x)$ has an asymptotic expansion if $|x^r R_r(x)| \to 0$ as $|x| \to \infty$ in $|\arg x| \leq t \leq \pi$. Ideally, we would like $|R_r(x)|$ to be bounded by the term $|\phi(x) A_{r+1}/x^{r+1}|$ in some sector; this corresponds to the well-known rule that the error committed in using an asymptotic series is less numerically than the first term omitted. The normal probability integral ratio belongs to this class. If $x \in N(0, 1)$ and $g(\cdot)$ is the density function, and if

$$R(t) = \Pr(x > t)/g(t) \quad (t > 0)$$

$$= t \int_{-\infty}^{\infty} \frac{g(x)dx}{x^2 + t^2},$$

$$= \frac{1}{t} - \frac{1}{t^3} + \frac{1.3}{t^5} - \ldots + (-1)^s \frac{1.3 \ldots (2s-1)}{t^{2s+1}} + R_s(t), \quad (1.17)$$

then

$$R_s(t) = t^{-(2s+1)} \int_{-\infty}^{\infty} \frac{x^{2s+2} g(x) dx}{x^2 + t^2}$$

$$< \frac{1.3.5 \ldots (2s+1)}{t^{2s+3}}.$$

Thus the truncated series in (1.17), if continued, would diverge, but the error in using a specified number of terms is less than the numerical value of the first term omitted; for a given t, we can at best find an optimal value of $R(t)$ from the series.

Stirling's expansion for the gamma function (Whittaker and Watson, 1915, p. 246),

$$\ln \Gamma(x) \sim (x - \tfrac{1}{2}) \ln x - x + \tfrac{1}{2} \ln (2\pi) + \phi(x),$$

where

$$\phi(x) = \sum_{r=1}^{\infty} \frac{(-1)^{r-1} B_r}{2r(2r-1) x^{2r-1}},$$

with $B_1 = \tfrac{1}{6}$, $B_2 = \tfrac{1}{30}$, $B_3 = \tfrac{1}{42}$, $B_4 = \tfrac{1}{30}$, etc. Bernoullian numbers, is an asymptotic expansion valid for $|\arg x| \leq \tfrac{1}{2}\pi - \delta$, with $\tfrac{1}{4}\pi > \delta > 0$. When $x > 0$, the error is less than the numerical value of the first term in $\phi(x)$ omitted. Accuracy of any desired degree can be arranged by using the recurrence $\Gamma(x+1) = x\Gamma(x)$, and evaluating $\Gamma(x + N)$ by the series, N being a positive integer.

Sharp asymptotic results rarely occur in statistics, though they do in mathematical analysis. Most applied mathematicians have their own choices of asymptotic delights; we like to recall the Hardy–Ramanujan (1918) asymptotic formula for $p(n)$, the number of unrestricted partitions of n, namely

$$p(n) = \sum_{1}^{\infty} P_s(n), \tag{1.18}$$

where

$$P_s(n) = L_s(n)\,\phi_s(n),$$

$$\phi_s(n) = \frac{1}{2\pi}\frac{s}{2}\frac{d}{dn}\frac{e^{km/s}}{m}$$

$$(m = \sqrt{(n - \tfrac{1}{24})})$$

$$L_s(n) = \sum_p w_{p,s}\,e^{-2in p\pi/s},$$

with p running through integers less than and prime to s, $w_{p,s}$ being a 24th root of unity (Hardy, 1940).

It turned out that 8 terms of (1.18) gave the value of $p(200)$ to within 0·004 of a thirteen-digit integer; it should however be remarked that this was an *a posteriori* error assessment, the authors having been fortunate in enlisting the aid of a local combinatorial genius to work out the answer with "pen and paper".

Very few attempts have been made to develop a general theory of asymptotic methods. The only one we know of is that due to Kruskal (1962), who *sets up several principles of asymptotology* such as (a) simplification, (b) interpretation, (c) wild behaviour, and (d) annihilation. Kruskal's interests are mainly, so it seems, in applications in theoretical physics.

There are only three modern textbooks: Erdélyi (1956) and De Bruijn (1961) and a treatment of classical asymptotics by Jeffreys (1962). Jeffreys in particular remarks on the uses of asymptotics as numerical tools, as opposed to their *representational* role in analysis; an account of this aspect and other theoretical development is given in Hardy (1949).

1.3.6 Asymptotic moments

The Cramér–Rao bound is a lower bound to the variance (under certain regularity conditions), *and for estimators in general gives no guarantee whatever that the variance even exists in finite samples.* Non-mathematical researchers may find this concept difficult to accept. What happens in producing an infinite variance (or other moment for that matter) is that there is a contribution from a "small" sub-space of the sample space which increases with N. In some cases this phenomenon is explicit; for example, the density of a Cauchy variate decreases for increasing X only as fast as X^{-2}. In this case there is no asymptotic expression at all for the variance. But suppose $\theta = 1/p$ is to be estimated for a sample from the binomial probability function whose pgf is $(pt + q)^n$; then $n/\bar{x}$ is the moment and ml estimator, and for $N \to \infty$,

$$E\frac{n}{\bar{x}} \sim \frac{1}{p}\left\{1 + \frac{\mu_2(\bar{x})}{n^2 p^2} - \frac{\mu_3(\bar{x})}{n^3 p^3} + \ldots\right\}$$

$$\sim \frac{1}{p}\left\{1 + \frac{q}{nNp} + \frac{q(q+1)}{n^2 N^2 p^2} + \ldots\right\};$$

however, the mean only exists if the probability of a zero value of $\bar{x}$ is ignored, a probability which tends to zero as $N \to \infty$.

1.3.7 Illustrations of asymptotic expansions in statistics

(a) **σ in normal sampling** Consider the ml estimators of μ, σ in sampling from $N(\mu, \sigma^2)$. Then

$$\hat{\sigma} = \left\{\frac{\Sigma(x_j - \bar{x})^2}{N}\right\}^{1/2},$$

$$= \sqrt{m_2}.$$

Basic asymptotics to the mean and variance of $\hat{\sigma}/\sigma$ would be unity and $\frac{1}{2}/N$. The latter is derived by the usual *incremental approach*; thus

$$\delta\hat{\sigma} \sim \frac{\delta m_2}{2\sigma}$$

and so

$$\text{Var } \hat{\sigma} \sim \frac{\text{Var } m_2}{4\sigma^2}$$

$$\sim \frac{\mu_4 - \mu_2^2}{4N\sigma^2}$$

$$= \sigma^2/2N$$

for the normal.

But the exact distribution of m_2 in normal sampling is known, and in fact

$$E(\hat{\sigma}) = \frac{N-1}{\sqrt{(2N)}} \frac{\Gamma(\tfrac{1}{2}N)\sigma}{\Gamma(\tfrac{1}{2}N + \tfrac{1}{2})}$$

$$\text{Var } \hat{\sigma} = \frac{N-1}{N}\sigma^2 - (E\hat{\sigma})^2,$$

with the asymptotic expansions

$$\left.\begin{aligned}
E(\hat{\sigma}/\sigma) \sim {} & 1 - 0{\cdot}75/N - 0{\cdot}21875/N^2 - 0{\cdot}070313/N^3 \\
& + 0{\cdot}028809/N^4 + 0{\cdot}058960/N^5 - 0{\cdot}035446/N^6 \\
& - 0{\cdot}163273/N^7 + 0{\cdot}110140/N^8 + 0{\cdot}895718/N^9 \\
& - 0{\cdot}633456/N^{10} - 8{\cdot}045846/N^{11} + 5{\cdot}819408/N^{12} \\
& + 106{\cdot}767006/N^{13} - 78{\cdot}148242/N^{14} - 1959{\cdot}839822/N^{15},
\end{aligned}\right\} \quad (1.19a)$$

and
$$\text{Var}\left(\frac{\hat{\sigma}}{\sigma}\right) = 1 - 1/N - \left(E\frac{\hat{\sigma}}{\sigma}\right)^2 \sim 0.5/N - 0.125/N^2 - 0.1875/N^3 + \ldots \tag{1.19b}$$

(This example has been used by the authors as a check on the recursive scheme for moments of sampling moments given by Shenton *et al.* (1971). Rounding-off error, or other sources, lead to a steady loss of accuracy, the higher the order of term considered. However, for $E(\hat{\sigma}/\sigma)$ there is 15-digit agreement for terms up to N^{-8}, and at N^{-12} the agreement is down to the leading first three digits.)

The higher-order terms clearly give much useful further information. Suppose we base our assessments on terms up to order N^{-2}. Then, for $N = 2$,

$$E(\hat{\sigma}/\sigma) \sim 1 - 0.375 - 0.055 = 0.570$$

$$\text{Var}(\hat{\sigma}/\sigma) \sim 0.250 - 0.031 = 0.219,$$

against the true values 0·5642 and 0·1817. If we truncate the series at the smallest numerical term (N^{-4} for the mean, N^{-8} for the variance) then the assessments are 0·5646 and 0·1820 with quite small errors. The first-order asymptotics 1 and $0.5/N$ give not the slightest hint that sample sizes as small as 2 might be acceptable.

Before leaving this example, note that in both series the coefficients first decrease and ultimately increase (or become oscillatory).

Example 1.11 Derive the series for $E(\hat{\sigma}/\sigma)$ given by (1.19a) by writing

$$E(\hat{\sigma}/\sigma) = \exp(a_1/N + a_2/N^2 + \ldots),$$

and using an asymptotic expansion for $\ln \Gamma(\cdot)$. Show that

$$a_s = -\frac{1}{s} - \frac{(-1)^s}{2(s+1)} + (-1)^s \sum_{r=1}^{[s/2]} \frac{2^{2r-1}(s-1)! B_{2r}}{(2r)!(s-2r+1)!}, \quad (s = 1, 2, \ldots)$$

where $[s]$ is the greatest integer $\leq s$, and $B_2 = \frac{1}{6}$, $B_4 = -\frac{1}{30}$, etc. are Bernoulli numbers.

If $E(\hat{\sigma}/\sigma) = A_0 + A_1/N + \ldots$, show that

$$s A_s = \sum_{r=1}^{s} r a_r A_{s-r}$$

and hence check the first few terms (or all) of (1.19b).

(b) *Fisher's linkage problem* Fisher (1925–) discussed estimators of θ in what amounts to a four-category structure with probabilities $(2 + \theta)/4$, $(1 - \theta)/4$, $(1 - \theta)/4$, $\theta/4$, based on observed frequencies a, b, c, d in a sample N. This problem belongs to class (a) of *1.3.4*, and $\hat{\theta}$ is an explicit function of the statistics $t_1 = (a - 2b - 2c - d)/N$, $t_2 = d/N$; in fact,

$$\hat{\theta} = \tfrac{1}{2}(t_1 + \sqrt{(t_1^2 + 8t_2)})/N, \quad d \neq 0, d = 0 \text{ and } a > 2b + 2c,$$
$$= 0, \quad d = 0 \text{ and } a \leq 2b + 2c.$$

The series for $\hat{\theta}$ is

$$\hat{\theta} = \theta + \tfrac{1}{2}(x - \theta - \tfrac{1}{2}) + \tfrac{1}{2}(\theta + \tfrac{1}{2}) \sum_{s=0}^{\infty} \frac{a_s Z^s}{(\theta + \tfrac{1}{2})^{2s}}, \qquad (1.20)$$

where
(i) $a_0, a_1, \ldots,$ are defined by
$$\sqrt{(1 + \omega)} = a_0 + a_1 \omega + a_2 \omega^2 + \ldots,$$
(ii) $x = t_1 - \zeta_1, \quad \zeta_1 = \theta - \tfrac{1}{2},$
(iii) $Z = (2\theta - 1)x + 8y + x^2,$
(iv) $y = t_2 - \xi_2, \quad \xi_2 = \tfrac{1}{4}\theta.$

The question of the convergence of (1.20) is of secondary importance; the real issue is the nature of the series for $E \hat{\theta}$.

The bivariate moment-generating function for x, y is

$$E\, e^{\alpha x + \beta y} = e^{-\alpha(\theta - 1/2) - \beta \theta/4} (p_1 e^a + p_2 e^{-2a} + p_3 e^{b-a})^N,$$

where $a = \alpha/N$, $b = \beta/N$, $p_1 = (2 + \theta)/4$, $p_2 = (1 - \theta)/2$, $p_3 = \theta/4$. To evaluate the moments $v_{r,s} = E\, x^r y^s$, projecting toward digital computerization, *the strategy is to derive results for $v_{r,0}$, $v_{0,s}$ and to use these to evaluate $v_{r,s}$ at internal lattice points.* We find:

$$v_{r+1,0} = -\sum_{s=0}^{r-1} \binom{r}{s} (A_{r-s}\, v_{s+1,0} - B_{r-s}\, v_{s,0}) \qquad (1.21)$$

$$(v_{0,0} = 1,\, v_{1,0} = 0;\, r = 1, 2, \ldots)$$

where

$$N^s A_s = (2 + \theta) 2^{s-2} + \tfrac{1}{2}(1 - \theta)(-1)^s, \quad s = 1, 2, \ldots$$
$$A_0 = 1,$$
$$8N^s B_s = (2 + \theta)(3 - 2\theta) 2^s - 2(1 - \theta)(3 + 2\theta)(-1)^s - \theta(1 + 2\theta)\delta_{s,0}$$
$$\delta_{r,s} = 0 \quad \text{if } r \neq s,\, \delta_{r,r} = 1.$$

From (1.21) the moments $v_{0,r+1}$ can be found replacing A_{r-s} by p and B_{r-s} by pq, where $n^{r-s}p = \theta/4$ and $q = 1 - \theta/4$. For internal points of the (x, y) space, use the bivariate recurrence

$$v_{r+1,s} = \sum_{\lambda=0}^{r} \sum_{\mu=0}^{s} \binom{r}{\lambda}\binom{s}{\mu} \{a_{\lambda,\mu}\, v_{r-\lambda,s-\mu} - b_{\lambda,\mu}\, v_{r+1-\lambda,s-\mu}\},$$

where

$$N^\mu a_{\lambda,\mu} = B_\lambda \quad \text{if } \mu = 0,$$
$$= -\theta(1 + 2\theta)/8 \quad \text{if } \lambda = 0, \mu = 1, 2, \ldots,$$
$$= 0, \text{ otherwise};$$
$$N^\mu b_{\lambda,\mu} = A_\lambda \quad \text{if } \mu = 0,$$
$$= \theta/4 \quad \text{if } \lambda = 0, \mu = 1, 2, \ldots,$$
$$= 0, \text{ otherwise}.$$

The mean value of $\hat{\theta}$ can now be programmed, as also $E(\hat{\theta} - \theta)^2 = E\{(\hat{\theta} - \theta) * (\hat{\theta} - \theta)\}$ and so on; and by applying the usual correction formula for non-central to central moments, series for the moments can be derived.

As illustrations we quote (Bowman and Shenton, 1971):

$$\theta = 0.1$$

$$E(\hat{\theta} - \theta) \sim -\frac{2\cdot 7}{N^2} - \frac{8\cdot 9^1}{N^3} - \frac{3\cdot 9^3}{N^4} - \frac{2\cdot 4^5}{N^5} - \frac{1\cdot 9^7}{N^6} - \frac{1\cdot 9^9}{N^7} - \frac{2\cdot 1^{11}}{N^8} + \frac{2\cdot 8^{13}}{N^9}$$

$$\text{Var } \hat{\theta} \sim \frac{3\cdot 2^{-1}}{N} + \frac{1\cdot 6}{N^2} + \frac{2\cdot 0^1}{N^3} + \frac{8\cdot 3^2}{N^4} + \frac{5\cdot 3^4}{N^5} + \frac{4\cdot 2^6}{N^6} + \frac{4\cdot 0^8}{N^7} + \frac{4\cdot 5^{10}}{N^8}$$

$$\mu_3(\hat{\theta}) \sim \frac{7\cdot 6^{-1}}{N^2} + \frac{4\cdot 2}{N^3} - \frac{5\cdot 0^1}{N^4} - \frac{4\cdot 2^3}{N^5} - \frac{3\cdot 5^5}{N^6} - \frac{3\cdot 6^7}{N^7} - \frac{4\cdot 0^9}{N^8} - \frac{4\cdot 9^{11}}{N^9}$$

$$\mu_4(\hat{\theta}) \sim \frac{3\cdot 0^{-1}}{N^2} + \frac{5\cdot 3}{N^3} + \frac{4\cdot 6^1}{N^4} + \frac{8\cdot 4^2}{N^5} + \frac{7\cdot 1^4}{N^6} + \frac{5\cdot 9^6}{N^7} + \frac{5\cdot 1^8}{N^8} + \frac{4\cdot 5^{10}}{N^9}$$

$$\theta = 0.9$$

$$E(\hat{\theta} - \theta) \sim -\frac{5\cdot 0^{-2}}{N^2} - \frac{2\cdot 5^{-1}}{N^3} - \frac{1\cdot 4}{N^4} - \frac{9\cdot 2}{N^5} - \frac{7\cdot 8^1}{N^6} - \frac{8\cdot 1^2}{N^7} - \frac{9\cdot 8^3}{N^8} - \frac{1\cdot 4^5}{N^9}$$

$$\text{Var } \hat{\theta} \sim \frac{1\cdot 9^{-1}}{N} + \frac{7\cdot 6^{-2}}{N^2} + \frac{4\cdot 0^{-1}}{N^3} + \frac{2\cdot 2}{N^4} + \frac{1\cdot 5^1}{N^5} + \frac{1\cdot 3^2}{N^6} + \frac{1\cdot 4^3}{N^7} + \frac{1\cdot 7^4}{N^8} + \frac{2\cdot 4^5}{N^9}$$

$$\mu_3(\hat{\theta}) \sim -\frac{3\cdot 2^{-1}}{N^2} - \frac{4\cdot 9^{-1}}{N^3} - \frac{2\cdot 4}{N^4} - \frac{1\cdot 7^1}{N^5} - \frac{1\cdot 5^2}{N^6} - \frac{1\cdot 6^3}{N^7} - \frac{2\cdot 0^4}{N^8} - \frac{2\cdot 9^5}{N^9}$$

$$\mu_4(\hat{\theta}) \sim \frac{1\cdot 0^{-1}}{N^2} + \frac{5\cdot 9^{-1}}{N^3} + \frac{2\cdot 7}{N^4} + \frac{1\cdot 8^1}{N^5} + \frac{1\cdot 6^2}{N^6} + \frac{1\cdot 7^3}{N^7} + \frac{2\cdot 2^4}{N^8} + \frac{3\cdot 2^5}{N^9}.$$

(Values were computed on IBM 360 model 91, using double precision floating-point arithmetic, the output of 8 significant digits being rounded off to 2 for brevity. Superscripts give the power of ten by which the entries are to be multiplied.)

Readers may like to use the entries for the moments given in Table 1.5 to assess (or guess, if they are experimentalists) the exact values evaluated by the *sample configuration method*.

Table 1.5

θ	N	μ_1	σ	μ_3	μ_4	$\sqrt{\beta_1}$	β_2
0·1	50	0·0983	0·0835	0·3924	0·1570	0·675	3·232
	75	0·0992	0·0674	0·1533	0·0653	0·500	3·158
	100	0·0996	0·0578	0·0815	0·0352	0·421	3·152
0·9	20	0·8998	0·0979	−0·8910	0·3623	−0·950	3·944
	25	0·8999	0·0873	−0·5552	0·2144	−0·836	3·699
	30	0·8999	0·0795	−0·3796	0·1420	−0·756	3·558
	35	0·8999	0·0735	−0·2761	0·1011	−0·696	3·465
	40	0·9000	0·0687	−0·2099	0·0756	−0·648	3·400
	50	0·9000	0·0613	−0·1331	0·0469	−0·577	3·312
	75	0·9000	0·0500	−0·0585	0·0200	−0·467	3·202
	100	0·9000	0·0433	−0·0327	0·0110	−0·404	3·150

$(\sqrt{\beta_1} = \mu_3/\mu_2^{3/2}, \beta_2 = \mu_4/\mu_2^2)$

1.3.8 Asymptotics in general

It is outside our scope here to attempt anything like a general treatment of the problem of statistical asymptotics — one may appreciate the scope of the subject by reference to the titles appearing in the Prague International Conference on this topic (September, 1973). However, one should mention studies relating to asymptotic moments, and the attempts to approximate theoretical distributions by equimomental models. Tschuprow (1918, 1920) discussed such things as the first eight central moments of the sample mean and mean values of products such as $(m'_r - \mu'_r)(m'_s - \mu'_s)(m'_t - \mu'_t)(m'_u - \mu'_u)$. Simplifying techniques for sampling moments were introduced by Thiele (1903), Craig (1928), and Fisher (1928). David (1949) showed how to shorten the algebra in the use of Fisher's k-statistics, with the properties of the coefficient of variation as an illustration. David and Johnson (1951), using asymptotic moments essentially, discussed the effects of non-normality on the power function of the F-test. Derksen (1939) and Wallace (1958) have made mainly theoretical studies of asymptotic approximations to distributions and moments.

Nearer to our present theme, we mention E.S. Pearson's (1930) studies of the distribution of $\sqrt{b_1}$, b_2 (skewness and kurtosis statistics) in normal sampling. He gives, for example, expressions for the standard deviation of $\sqrt{b_1}$ to order $N^{-7/2}$, $\beta_2(\sqrt{b_1})$ to N^{-3}; thus

$$\sigma(\sqrt{b_1}) \sim \sqrt{(6/N)} \left\{ 1 - \frac{3}{N} + \frac{6}{N^2} - \frac{15}{N^2} \right\}, \tag{1.22a}$$

$$\beta_2(\sqrt{b_1}) \sim 3 + \frac{36}{N} - \frac{864}{N^2} + \frac{12096}{N^3}. \tag{1.22b}$$

Pearson considers it safe, from the point of view of convergence, to use (1.22a) for $N \geq 50$ and (1.22b) for $N \geq 100$ (and possibly $N = 75$). Actually, he gives

$N = 50$: $\sigma(\sqrt{b_1}) \sim 0.3464(1 - 0.060000 + 0.002400 - 0.000120)$

$\qquad\qquad = 0.3264$

$N = 75$: $\beta_2(\sqrt{b_1}) \sim 3 + 0.480000 - 0.153600 + 0.028668$

$\qquad\qquad = 3.3551$

$N = 100$: $\beta_2(\sqrt{b_1}) \sim 3 + 0.360000 - 0.086400 + 0.012096$

$\qquad\qquad = 3.2857$

The correct values are 0.3264, 3.3511 and 3.2844, showing highly satisfactory agreement. Actually the exact values of $\sigma^2(\sqrt{b_1})$ and $\beta_2(\sqrt{b_1})$ are rational fractions in N, and (1.22a) and (1.22b) converge for $N > 3$ and $N > 9$ respectively.

REFERENCES

1. Aitken, A.C. and Silverstone, H. (1942). "On the estimation of statistical parameters". *Proc. Roy. Soc. Edinb.*, **61**, 186–94.
2. Anscombe, F.J. (1948). "The transformation of Poisson, binomial and negative binomial data". *Biometrika*, **35**, 246–54.
3. Antle, C.E. and Bain, L.J. (1969). "A property of maximum likelihood estimators of location and scale parameters". *SIAM Rev.*, **11**, No. 2, 251–3.
4. Bartlett, M.S. (1936). "Square root transformation in analysis of variance". Supp. *J. Roy. Statist. Soc.*, **3**, 68–78.
5. Bowman, K.O. and Shenton, L.R. (1971). "Small sample properties of the maximum likelihood estimator associated with Fisher's linkage problem". *J. Statist. Comput. Simul.*, **1**, 71–85.
6. Brillinger, D.R. (1964). "The asymptotic behaviour of Tukey's general method of setting approximate confidence limits (the Jacknife) when applied to maximum likelihood estimates". *Revue de L'Institut International de Statistique*, **32**, 202–6.
7. Cox, D.R. and Hinkley, D.V. (1974). *Theoretical Statistics*. Chapman & Hall, London.
8. Craig, C.C. (1928). "An application of Thiele's semi-invariants to the sampling problem". *Metron*, **7**, No. 3, 3–74.
9. Cramér, H. (1946). *Mathematical Methods of Statistics*. Princeton University Press.
10. David, F.N. (1949). "Note on the application of Fisher's *k*-statistics". *Biometrika*, **36**, 383–93.
11. David, F.N. and Johnson, N.L. (1951). "The effect of non-normality on the power function of the *F*-test in the analysis of variance". *Biometrika*, **38**, 43–57.

12. de Bruijn, N.G. (1961). *Asymptotic Methods in Analysis*. North-Holland Publishing Co., Amsterdam.
13. Derksen, J.B.D. (1939). "On some infinite series introduced by Tschuprow". *Ann. Math. Statist.*, **10**, 380–3.
14. Edwards, A.W.F. (1974). "The history of likelihood". *Int. Stat. Rev.*, **42**, 9–15.
15. Erdélyi, A. (1956). *Asymptotic Expansions*. Dover Publications, New York.
16. Feller, W. (1966). *An Introduction to Probability Theory and its Applications*, Vol. 2 (2nd edn). John Wiley & Sons, New York.
17. Fisher, R.A. (1921). "On the mathematical foundations of theoretical statistics". *Phil. Trans.*, *A*, **222**, 308–68.
18. Fisher, R.A. (1925). "Theory of statistical estimation". *Proc. Camb. Phil. Soc.*, **22**, 700–25.
19. Fisher, R.A. (1925–). *Statistical Methods for Research Workers* (14th edn, 1970), Oliver and Boyd, Edinburgh.
20. Fisher, R.A. (1928). "Moments and product-moments of sampling distributions". *Proc. Lond. Math. Soc.*, **30**, 199–238.
21. Fisher, R.A.,(1941). "The negative binomial distribution". *Ann. Eugen.*, **11**, 182–7.
22. Freeman, H. (1963). *Introduction to Statistical Inference*. Addison-Wesley Publishing Co., Reading, Mass.
23. Geary, R.C. (1942). "The estimation of many parameters". *J. Roy. Statist. Soc.*, **105**, 213–17.
24. Godwin, H.J. (1964). *Inequalities on Distribution Functions*. Charles Griffin & Co., London & High Wycombe; Hafner Press, New York.
25. Gray, H.L. and Schucany, W.R. (1972). *The Generalized Jacknife Statistic*. Marcel Dekker, New York.
26. Haldane, J.B.S. (1953). "The estimation of two parameters from a sample". *Sankhyā*, **12**, 313–20.
27. Haldane, J.B.S. and Smith, Sheila M. (1956). "The sampling distribution of a maximum likelihood estimate". *Biometrika*, **43**, 96–103.
28. Hardy, G.H. (1940). *Ramanujan: Twelve Lectures on Subjects suggested by his Life and Work*. Chelsea Publishing Co., New York.
29. Hardy, G.H. (1949). *Divergent Series*. Clarendon Press, Oxford.
30. Huzurbazar, V.S. (1948). "The likelihood equation, consistency and the maxima of the likelihood function". *Ann. Eugen.*, **14**, 185–200.
31. Jeffreys, H. (1962). *Asymptotic Expansions*. Oxford, Clarendon Press.
32. Katti, S.K. and Gurland, J. (1962). "Efficiency of certain methods of estimation for the negative binomial and the Neyman Type A distributions". *Biometrika*, **49**, 215–26.
33. Kendall, M.G. and Stuart, A. (1969). *The Advanced Theory of Statistics*, Vol. 1, *Distribution Theory* (3rd edn). Charles Griffin & Co., London & High Wycombe; Hafner Press, New York.
34. Kendall, M.G. and Stuart, A. (1973). *The Advanced Theory of Statistics*, Vol. 2, *Inference and Relationship* (3rd edn). Charles Griffin and Co., London & High Wycombe; Hafner Press, New York.
35. Kruskal, M.D. (1962). *Asymptotology: Mathematical Models in the Physical Sciences*. Prentice Hall Inc., Englewood Cliffs, N.J.
36. Norden, R.H. (1972). "A survey of maximum likelihood estimation". *Int. Stat. Rev.*, **40**, No. 3, 329–54.
37. Norden, R.H. (1973). "A survey of maximum likelihood estimation, Part 2". *Int. Stat. Rev.*, **41**, No. 1, 39–58.
38. Pearson, E.S. (1930). "A further development of tests for normality". *Biometrika*, **22**, 239–49.
39. Pearson, E.S. (1974). "Memories of the impact of Fisher's work in the 1920's". *Int. Stat. Rev.*, **42**, No. 1, 5–8.

40. Quenouille, M.H. (1956). "Notes on bias in estimation". *Biometrika*, **43**, 353–60.
41. Rao, C.R. (1952). *Advanced Statistical Methods in Biometric Research* John Wiley & Sons, New York.
42. Rao, C.R. (1962). "Apparent anomalies and irregularities in maximum likelihood estimation". *Sankhyā*, Ser. A, **24**, Pt. 1, 73–101.
43. Rao, C.R. (1965). *Linear Statistical Inference and its Applications.* John Wiley and Sons, New York.
44. Shenton, L.R. (1970). "The so-called Cramér–Rao inequality". Letter to the Editor, *American Statistician*, **24**, No. 2, 36.
45. Shenton, L.R., Bowman, K.O., and Sheehan, D. (1971). "Sampling moments of moments associated with univariate distributions". *J. Roy. Statist. Soc. (B)*, **33**, No. 3, 444–57.
46. Tschuprow (Tchouproff), Al. A. (1918). "On the mathematical expectation of the moments of frequency distributions". *Biometrika*, **12**, 140–69; **13**, 283–95.
47. Thiele, T.N. (1903). *The Theory of Observations.* C. & E. Layton, London.
48. Tukey, J.W. (1958). "Bias and confidence in not quite large samples" (Abstract). *Ann. Math. Statist.*, **29**, 614.
49. Wallace, D.L. (1958). "Asymptotic approximations to distributions". *Ann. Math. Statist.*, **29**, 635–54.
50. Whittaker, E.T. and Watson, G.N. (1915). *A Course of Modern Analysis.* Cambridge Univ. Press.

2 SINGLE PARAMETER ESTIMATION[*]

2.1 The problem

We assume that a population consists of a denumerable set of classes, there being n_j relative observations in the jth class for a sample of N, with $E\, n_j = p_j(\theta)$, where p_j is a function of a single parameter θ. The log likelihood is

$$L(n, \theta) = N \Sigma\, n_j \log p_j(\theta) \qquad (2.1)$$
$$= N(n_j \Gamma^j), \quad \text{say},$$

summed over classes; where there is little risk of ambiguity, $(n\Gamma)$ will be used for $(n_j\, \Gamma^j)$. For the maximum likelihood estimator (mle) $\hat{\theta}$, writing L_1 for the derivative,

$$L_1(n, \hat{\theta}) = 0, \qquad (2.2)$$

or, using a Taylor-series approximation,

$$(n\Gamma_1) + \frac{x}{1!}(n\Gamma_2) + \frac{x^2}{2!}(n\Gamma_3) + \ldots = 0, \qquad (2.3)$$

where $x = \hat{\theta} - \theta$, $\Gamma_s = d^s \log p/d\theta^s$.

2.2 Methods available for moment evaluation

There are three possibilities which can be used to evaluate the non-central moments of x, and thus using the usual correction formulae, the central moments. The main concern is to discover which methods can be exploited in the single-parameter case, and which methods have the potential for generalization to the multiparameter case. It is the latter situation which seems to introduce grave complexities, to the extent that even N^{-2} terms in the covariances are far from simple structures.

The first method is to use Lagrange's expansion on equation (2.3), noting

---

[*] The reader who does not need to understand the detailed derivation of the formulae and who is interested mainly in applications may skip portions of this chapter. However, for quick appreciation, refer to sections: **2.1** for a statement of the problem; **2.2** for possible approaches to solutions; **2.3** for basic statistics and their expectations; **2.7** for the Taylor-series approach, and in particular paragraphs *2.7.4* for details of the bias and *2.7.5–2.7.7* for detailed formulae for four moments of a maximum likelihood estimator using logarithmic derivatives; and section **2.8** for detailed formulae using derivatives of the probability function.

that $-\mathrm{E}(n\Gamma_2) = -(p\Gamma_2) > 0$ in general and is Fisher's information on θ supplied by samples of N. Haldane and Smith (1956, p. 99) recognized that Lagrange's expansion was involved, at least by name, and left it at that.

The second approach is what we describe as an adjusted order of magnitude process. Equation (2.3), and powers of it, and adjusted forms of (2.3) found by using random multipliers, are manipulated to yield a particular coefficient of N^{-s} ($s = 1, 2, \ldots$) in the expectation. The method was inspired by a device used by Cox and Snell (1968) to find the N^{-1} bias and covariance in ml estimators associated with residuals of regression models (see in particular pp. 251–2).

Method three is to regard $\hat{\theta}$ as a c-dimensional variate, the arguments being $(n_1, n_2, \ldots, n_c)$; although this Taylor-series approach seems very indirect, its structure is straightforward, and it has yielded a number of new results (Shenton and Bowman, 1963).

Each of these approaches requires, sooner or later, the evaluation of the mean values of products of linear forms, i.e. $\mathrm{E}\{(n\Gamma_1)^r (n\Gamma_2)^s \ldots\}$, which we now consider.

The reader should note that the sample size will be denoted by N for this and succeeding chapters.

2.3 Expectation of products of linear forms

A first version of this was given by Shenton (1959), and later was elaborated by Bowman (1963).

Let $n_j = p_j + \epsilon_j$, so that $\mathrm{E}\,\epsilon_j = 0$, and for an arbitrary non-null sequence $\{h\}$,

$$\mathrm{E}\,e^{\alpha(h\epsilon)} = e^{-\alpha(hp)}(p_j\,e^{\alpha h_j/N})^N \tag{2.4}$$

from statistical independence. For example, using $\mathrm{E}_m \phi$ to denote the N^{-m} coefficient in $\mathrm{E}\phi$,

$$\begin{aligned}
\mathrm{E}_1(h\epsilon)^2 &= (h^2 p) - (hp)^2, \\
\mathrm{E}_2(h\epsilon)^3 &= (h^3 p) - 3(h^2 p)(hp) + 2(hp)^3, \\
\mathrm{E}_2(h\epsilon)^4 &= 3\{(h^2 p) - (hp)^2\}^2.
\end{aligned} \tag{2.5a}$$

If $h = \Gamma_1$, then $(hp) = 0$ and many expressions simplify. Again, the operator $(i\,\partial/\partial h)$ commutates with the expectation operator, so that, for example, from (2.5a),

$$\mathrm{E}_1(h\epsilon)(i\epsilon) = (hip) - (hp)(ip)$$

$$\mathrm{E}_2(h\epsilon)^2(i\epsilon) = (h^2 ip) - 2(hip)(hp) - (h^2 p)(ip) + 2(hp)^2(ip). \tag{2.5b}$$

Similarly,

$$\mathrm{E}_2(h\epsilon)(i\epsilon)(j\epsilon) = (hijp) - (hp)(ijp) - (ip)(hjp) - (jp)(hip)$$
$$+ 2(hp)(ip)(jp). \tag{2.5c}$$

We have evaluated all products of three to order N^{-2}, all products of four to order N^{-3}, and the following individual cases (for brevity $E_s(1^r 2^s \ldots)$) refers to $E_s\{(h\epsilon)^r (i\epsilon)^s \ldots\}$): $E_3(1^5)$, $E_3(1^4 2)$, $E_3(1^3 23)$; $E_4(1^5)$, $E_4(1^4 2)$, $E_4(1^3 23)$; $E_3(1^6)$, $E_3(1^5 2)$, $E_3(1^4 23)$, $E_3(1^3 234)$; $E_4(1^6)$, $E_4(1^5 2)$, $E_4(1^4 23)$; $E_4(1^7)$, $E_4(1^6 2)$, $E_4(1^5 23)$, $E_4(1^4 234)$; $E_4(1^8)$, $E_4(1^7 2)$, $E_4(1^6 23)$, $E_4(1^5 234)$, $E_4(1^4 2345)$. (Listings are deposited in the Library of Virginia Polytechnic and State University, and also in the Library of the Royal Statistical Society.)

2.4 Order of magnitude of products of linear forms

From (2.4),

$$E\, e^{\alpha(h\epsilon)} = \left[e^{-\frac{\alpha(hp)}{N}} \left(1 + \frac{\alpha}{N}(ph) + \frac{\alpha^2}{2! N^2}(ph^2) + \ldots \right) \right]^N,$$

$$= \left(1 + \frac{\alpha^2}{2! N^2} H_2 + \frac{\alpha^3}{3! N^3} H_3 + \ldots \right)^N,$$

where $H_2, H_3, \ldots$ are independent of N. Evidently the highest power of N^{-1} in $E(h\epsilon)^r$ is $N^{-(r-1)}$. The lowest depends on whether r is even or odd. If $r = 2R$, then the lowest power arises from

$$\binom{N}{R} \left[\frac{\alpha^2}{2! N^2} H_2 + \ldots \right]^R$$

and is therefore N^{-R}. If $r = 2R + 1$, the lowest power arises from

$$\binom{N}{R} \left[\frac{\alpha^2}{2! N^2} H_2 + \frac{\alpha^3}{3! N^3} H_3 + \ldots \right]^R$$

and is therefore $N^{-(R+1)}$. Hence, if $[x]$ refers to the greatest integer less than x, then

$$E(h\epsilon)^r = \sum_{s=r^*}^{r-1} i_s / N^s \quad (r^* = [\tfrac{1}{2}(r+1)]).$$

Similarly the coefficients of powers of N^{-1} in the product $E\{(h\epsilon)^r (i\epsilon)^s \ldots (w\epsilon)^z\}$ range from $[\tfrac{1}{2}(S+1)]$ to $S - 1$, where $S = r + s + \ldots + z$.

2.5 Lagrange's expansion

Define $d_s = (n\Gamma_s)/(n\Gamma_2)$, $s = 1, 2, \ldots$, so that $d_2 = 1$ and the likelihood equation becomes

$$d_1 + \frac{x}{1!} d_2 + \frac{x^2}{2!} d_3 + \ldots = 0$$

or

$$d_1 + x\left\{1 + \frac{xd_3}{2!} + \frac{x^2 d_4}{3!} + \ldots\right\} = 0 \tag{2.6}$$

or

$$x = -d_1/\psi(x), \text{ say.}$$

The formal Lagrange expansion now gives, for a well-behaved function $f(\cdot)$,

$$f(x) = f(0) + \sum_{1}^{\infty} \frac{1}{s!} \frac{d^{s-1}}{dx^{s-1}} \left[\frac{-d_1}{1 + \frac{x}{2!}d_3 + \ldots}\right]^s f^{(1)}(x) \Bigg|_{x=0} \tag{2.7}$$

In particular,

$$\hat{\theta} = \theta + \sum_{s=1}^{\infty} d_1^s D_s, \tag{2.8}$$

where

$$D_1 = -1 \qquad D_2 = -d_3/2$$
$$D_3 = (d_4 - 3d_3^2)/6 \qquad D_4 = -(d_5 - 10d_3 d_4 + 15d_3^3)/24$$
$$D_5 = (d_6 - 10d_4^2 - 15d_3 d_5 + 105d_3^2 d_4 - 105d_3^4)/120$$
$$D_6 = -(d_7 - 21d_3 d_6 - 35d_4 d_5 + 210d_3^2 d_5 + 280d_3 d_4^2 - 1260d_3^3 d_4$$
$$+ 945d_3^5)/720,$$

and so on.

To find a central moment, we set up expansions for the required non-central moments and apply correction formulae. Consider, for example, the bias of $\hat{\theta}$. Since

$$d_1 = \{(p\Gamma_1) + (\epsilon\Gamma_1)\}/\{(p\Gamma_2) + (\epsilon\Gamma_2)\}$$

and

$$(p\Gamma_1) = \sum \frac{\partial p_j}{\partial \theta},$$

we see that if we may interchange the summation and differentiation operators (the corresponding expression in the continuous case would require the range of the variate to be θ-independent), then

$$d_1 = (\epsilon\Gamma_1)/\{(p\Gamma_2) + (\epsilon\Gamma_2)\}, \tag{2.9}$$

where $(p\Gamma_2) = -(p\Gamma_1^2) = -I$ (with the same proviso on operator commutability). Since, also, in expectation $(n\Gamma_s)$, $s = 2, 3, \ldots$, is of order 1, we see that orders of magnitude with respect to the sample size N in (2.8) or its powers are dominated by powers of d_1. Thus terms needed in E $\hat{\theta}$ to include the $N^{-1}, N^{-2}, N^{-3}, \ldots$ coefficients are those up to and including $d_1^2, d_1^4, d_1^6, \ldots$, respectively. We consider two examples (we assume throughout that the range of the variate is θ-independent, so that for example $(p\Gamma_2) = -(p\Gamma_1^2)$).

N^{-1} *term in the bias*

$$\begin{aligned}
\mathrm{E}_1 d_1 &= \mathrm{E}_1\left[\frac{(\epsilon\Gamma_1)}{-I}\left(1 + \frac{(\epsilon\Gamma_2)}{I} + \ldots\right)\right] \\
&= -\frac{1}{I^2}\mathrm{E}_1(\epsilon\Gamma_1)(\epsilon\Gamma_2) \\
&= -\frac{(p\Gamma_1\Gamma_2)}{I^2}
\end{aligned}$$

from (2.5b). Similarly

$$\begin{aligned}
\mathrm{E}_1(d_1^2 D_2) &= -\tfrac{1}{2}\mathrm{E}_1\left[\frac{(\epsilon\Gamma_1)^2\{(p\Gamma_3) + (\epsilon\Gamma_3)\}}{(-I + (\epsilon\Gamma_2))^3}\right] \\
&= \frac{(p\Gamma_3)}{2I^3}\mathrm{E}_1(\epsilon\Gamma_1)^2 \\
&= \frac{(p\Gamma_3)}{2I^2}.
\end{aligned}$$

Hence

$$\mathrm{E}_1\hat{\theta} = \{(p\Gamma_1\Gamma_2) + \tfrac{1}{2}(p\Gamma_3)\}/I^2, \tag{2.10}$$

as given by Bartlett (1952), Haldane (1953), Haldane and Smith (1956), Shenton (1959), Shenton and Wallington (1962), Shenton and Bowman (1963).

This method could be used for higher-order terms and moments, but it is not obvious how to extend it to the multiparameter case.

2.6 Adjusted order of magnitude method

2.6.1 Returning to (2.3), we define

$$\begin{aligned}
y_s &= (n\Gamma_s) \\
&= (p\Gamma_s) + (\epsilon\Gamma_s) \\
&= c_s + \gamma_s \text{ say.}
\end{aligned}$$

Now squaring the expression, we have

$$(c_2^2 + 2c_2\gamma_2 + \gamma_2^2)x^2 + (c_2 c_3 + c_2\gamma_3 + c_3\gamma_2 + \gamma_2\gamma_3)x^3 \\ + \left[\frac{(c_3 + \gamma_3)^2}{4} + \frac{(c_2 + \gamma_2)(c_4 + \gamma_4)}{3}\right]x^4 + \ldots = \gamma_1^2. \tag{2.11}$$

Isolating the N^{-1} terms,

$$c_2^2 E_1 x^2 = E_1 \gamma_1^2$$

or

$$E_1 x^2 = (p\Gamma_1^2)/I^2$$

$$= 1/I, \qquad (2.12a)$$

so that to order N^{-1},

$$\text{Var}_1 x = E_1 x^2$$

$$= 1/I. \qquad (2.12b)$$

2.6.2 Now turning to N^{-2} coefficients, we find

$$c_2^2 E_2 x^2 + 2c_2 E_2 \gamma_2 x^2 + E_2 \gamma_2^2 x^2 + c_2 c_3 E_2 x^3 + c_2 E_2 \gamma_3 x^3$$

$$+ c_3 E_2 \gamma_2 x^3 + \left(\frac{c_3^2}{4} + \frac{c_2 c_4}{3} \right) E_2 x^4 = E_2 \gamma_1^2. \qquad (2.13a)$$

There are seven unknowns. Evaluate these, starting with the highest powers of x and downwards. From (2.3), the fourth power

$$c_2^4 E_2 x^4 = E_2 \gamma_1^4. \qquad (2.13b)$$

From the third power,

$$(c_2^3 + 3c_2^2 \gamma_2 + \ldots)x^3 + \tfrac{3}{2}(c_2^2 c_3 + \ldots)x^4 + \ldots = -\gamma_1^3,$$

it follows, keeping in mind orders of magnitude,

$$c_2^3 E_2 x^3 + 3c_2^2 E_2 \gamma_2 x^3 + \tfrac{3}{2} c_2^2 c_3 E_2 x^4 = -E_2 \gamma_1^3 \qquad (2.13c)$$

$$c_2^3 E_2 \gamma_2 x^3 = -E_2 \gamma_2 \gamma_1^3 \qquad (2.13d)$$

$$c_2^3 E_2 \gamma_3 x^3 = -E_2 \gamma_3 \gamma_1^3. \qquad (2.13e)$$

Also, from the square of (2.3) given in (2.11),

$$c_2^2 E_2 \gamma_2^2 x^2 = E_2 \gamma_1^2 \gamma_2^2 \qquad (2.13f)$$

$$c_2^2 E_2 \gamma_2 x^2 + 2c_2 E_2 \gamma_2^2 x^2 + c_2 c_3 E_2 \gamma_2 x^3 = E_2 \gamma_2 \gamma_1^2. \qquad (2.13g)$$

Solving (2.13) for $E_2 x^2$, we can then evaluate $\text{Var}_2 x = E_2 x^2 - (E_1 x)^2$, in which the correction term has already been derived by our first method in (2.10). Using the second method on $E_1 x$, we have

$$(c_2 + \gamma_2)x + \tfrac{1}{2}(c_3 + \gamma_3)x^2 + \ldots = -\gamma_1$$

and

$$(c_2 + \gamma_2)\gamma_2 x + \ldots = -\gamma_1 \gamma_2,$$

so, taking expectations,

$$c_2 E_1 x + E_1 \gamma_2 x + \tfrac{1}{2} c_3 E_1 x^2 = 0$$

$$c_2 E_1 \gamma_2 x = -(p\Gamma_1\Gamma_2).$$

Solving for $E_1 x$, using (2.12a) for $E_1 x^2$ leads to (2.10) again. Finally we have, for the N^{-2} variance,

$$\text{Var}_2 \hat{\theta} = -1/I + \{3(p\Gamma_2^2) + 3(p\Gamma_1\Gamma_3) + 2(p\Gamma_1^2\Gamma_2) + (p\Gamma_4)\}/I^3$$
$$+ \{5(p\Gamma_1\Gamma_2)^2 + 11(p\Gamma_3)(p\Gamma_1\Gamma_2) + 7(p\Gamma_3)^2/2$$
$$+ (p\Gamma_3)(p\Gamma_1^3)\}/I^4. \tag{2.14}$$

Note in passing that some slight simplification would result from dividing throughout the equations by the coefficient of the first term in x; for example, (2.11) would be divided by $(c_2 + \gamma_2)^2$.

2.6.3 Again, consider the strategy for finding $E_2 x$. We start from (2.3). Clearly we have to include terms up to x^4. Contributions will arise from

(a) $c_5 x^4$;

(b) $x^3, x^3 \gamma_4$;

(c) $x^2, x^2 \gamma_3$;

(d) $x\gamma_2, x$.

For (a), fourth power (2.3). For (b), cube (2.3) and use it along with γ_4 as a multiplier; note that these equations will introduce terms in x^4. For (c) use (2.11) and a multiple by γ_3; note these equations will involve x^3 and x^4 terms. For (d) use (2.3) and the same equation with γ_2 as a multiplier; note these equations will involve contributions from x^2, x^3 and x^4 terms. Finally, solving for $E_2 x$, we have

$$E_2 \hat{\theta} = \sum_{s=2}^{5} a_s^{(1)} I^{-s}, \tag{2.15}$$

where

$$2a_2^{(1)} = -(p\Gamma_3) - 2(p\Gamma_1\Gamma_2),$$
$$8a_3^{(1)} = 4(p\Gamma_1^2\Gamma_3) + 8(p\Gamma_1\Gamma_2^2) + 4(p\Gamma_1\Gamma_4) + 12(p\Gamma_2\Gamma_3) + (p\Gamma_5),$$
$$12a_4^{(1)} = 2(p\Gamma_4)(p\Gamma_1^3) + 18(p\Gamma_3)(p\Gamma_1^2\Gamma_2) + 15(p\Gamma_3)(p\Gamma_4)$$
$$+ 36(p\Gamma_3)(p\Gamma_2^2) + 36(p\Gamma_3)(p\Gamma_1\Gamma_3) + 36(p\Gamma_1\Gamma_2)(p\Gamma_1\Gamma_3)$$
$$+ 36(p\Gamma_1\Gamma_2)(p\Gamma_2^2) + 24(p\Gamma_4)(p\Gamma_1\Gamma_2),$$
$$8a_5^{(1)} = 4(p\Gamma_3)^2(p\Gamma_1^3) + 15(p\Gamma_3)^3 + 48(p\Gamma_3)(p\Gamma_1\Gamma_2)^2$$
$$+ 60(p\Gamma_3)^2(p\Gamma_1\Gamma_2).$$

2.6.4 We shall not pursue this approach here. Clearly it is fairly straightforward, but has little advantage over the Lagrange expansion (2.7). It can, however, be used in the multiparameter case. We pass on to the h-variate Taylor approach, which was the initial attack on the problem (Shenton and Bowman, 1963).

2.7 c-variate Taylor approach

2.7.1 Since $\hat{\theta}$ is a function of $(n_1, n_2, \ldots, n_c)$, we have

$$\hat{\theta} = \theta + \Phi_1 + \frac{1}{2!}\Phi_2 + \frac{1}{3!}\Phi_3 + \ldots, \tag{2.16}$$

where

$$\Phi_1 = \epsilon_r \bar{\theta}_r, \quad \Phi_2 = \epsilon_r \epsilon_s \bar{\theta}_{rs}, \quad \Phi_3 = \epsilon_r \epsilon_s \epsilon_t \bar{\theta}_{rst},$$

etc., and

$$\bar{\theta}_r = \left.\frac{\partial \hat{\theta}}{\partial n_r}\right|_{\hat{\theta}=\theta, n=p}$$

$$= \hat{\theta}_r|_{\hat{\theta}=\theta, n=p}$$

and similar expressions for $\bar{\theta}_{rs}$, etc..

For brevity we use

$$\left.\begin{array}{l} J = (p\Gamma_3), \quad K = (p\Gamma_4) \\ L = (p\Gamma_5), \quad M = (p\Gamma_6), \end{array}\right\} \tag{2.17}$$

where, as before, $-I = (p\Gamma_2)$. Moreover we write $\hat{I} = -(n\hat{\Gamma}_2)$, $\hat{J} = (n\hat{\Gamma}_3)$, etc.

2.7.2 *Derivatives of* $\hat{\theta}$

Writing the likelihood equation

$$\Sigma n_r \hat{\Gamma}_1^r = 0, \tag{2.18}$$

we have, differentiating with respect to n_r,

$$\hat{\Gamma}_1^r - \hat{\theta}_r \hat{I} = 0, \tag{2.19a}$$

from which

$$\bar{\theta}_r = \Gamma_1^r/I. \tag{2.19b}$$

Differentiating (2.19a) with respect to n_s gives

$$\hat{\theta}_s \hat{\Gamma}_2^r + \hat{\theta}_r \hat{\Gamma}_2^s + \hat{\theta}_r \hat{\theta}_s \hat{J} - \hat{\theta}_{rs} \hat{I} = 0, \tag{2.20a}$$

from which, using (2.19b),

$$\bar{\theta}_{rs} = (\Gamma_1^r \Gamma_2^s + \Gamma_2^r \Gamma_1^s)/I^2 + J\Gamma_1^r \Gamma_1^s/I^3. \tag{2.20b}$$

Similarly,

$$\hat{\theta}_s\hat{\theta}_t\hat{\Gamma}_3^r + \hat{\theta}_r\hat{\theta}_t\hat{\Gamma}_3^s + \hat{\theta}_r\hat{\theta}_s\hat{\Gamma}_3^t + \hat{\theta}_{st}\hat{\Gamma}_2^r + \hat{\theta}_{rt}\hat{\Gamma}_2^s + \hat{\theta}_{rs}\hat{\Gamma}_2^t$$
$$+ \hat{J}(\hat{\theta}_{st}\hat{\theta}_r + \hat{\theta}_{rt}\hat{\theta}_s + \hat{\theta}_{rs}\hat{\theta}_t) + \hat{K}\hat{\theta}_r\hat{\theta}_s\hat{\theta}_t - \hat{I}\hat{\theta}_{rst} = 0, \quad (2.21a)$$

from which

$$\bar{\theta}_{rst} = (\Gamma_3^r \Gamma_1^s \Gamma_1^t + \Gamma_1^r \Gamma_3^s \Gamma_1^t + \Gamma_1^r \Gamma_1^s \Gamma_3^t)/I^3$$
$$+ 2(\Gamma_2^r \Gamma_2^s \Gamma_1^t + \Gamma_2^r \Gamma_1^s \Gamma_2^t + \Gamma_1^r \Gamma_2^s \Gamma_2^t)/I^3$$
$$+ (KI + 3J^2)\Gamma_1^r \Gamma_1^s \Gamma_1^t / I^5$$
$$+ 3J(\Gamma_2^r \Gamma_1^s \Gamma_1^t + \Gamma_1^r \Gamma_2^s \Gamma_1^t + \Gamma_1^r \Gamma_1^s \Gamma_2^t)/I^4. \quad (2.21b)$$

The expression for $\hat{\theta}_{rstu}$ involves 35 terms and may be written

$$\sum_4 \hat{\Gamma}_4^r \hat{\theta}_s \hat{\theta}_t \hat{\theta}_u + \hat{K} \sum_6 \hat{\theta}_{rs} \hat{\theta}_t \hat{\theta}_u + \hat{J} \sum_3 \hat{\theta}_{rs} \hat{\theta}_{tu} + \sum_{12} \hat{\Gamma}_3^r \hat{\theta}_s \hat{\theta}_{tu}$$
$$+ \hat{J} \sum_4 \hat{\theta}_r \hat{\theta}_{stu} + \sum_4 \hat{\Gamma}_2^r \hat{\theta}_{stu} + \hat{L}\hat{\theta}_r\hat{\theta}_s\hat{\theta}_t\hat{\theta}_u - \hat{I}\hat{\theta}_{rstu} = 0, \quad (2.22)$$

where a summation term involves the various combinations of selections from r, s, t, u, the number of terms being indicated under the sigma symbol. From this, using (2.20b) and (2.21b), the expression for $\bar{\theta}_{rstu}$ follows.

For the fifth-order derivative we have

$$I\bar{\theta}_{rstuv} = \sum_5 \Gamma_5^r \bar{\theta}_s \bar{\theta}_t \bar{\theta}_u \bar{\theta}_v + \sum_{30} \Gamma_4^r \bar{\theta}_{st} \bar{\theta}_u \bar{\theta}_v + L \sum_{10} \bar{\theta}_{rs} \bar{\theta}_t \bar{\theta}_u \bar{\theta}_v$$
$$+ K \sum_{10} \bar{\theta}_{rst} \bar{\theta}_u \bar{\theta}_v + K \sum_{15} \bar{\theta}_{rs} \bar{\theta}_{tu} \bar{\theta}_v + J \sum_{10} \bar{\theta}_{rst} \bar{\theta}_{uv}$$
$$+ \sum_{15} \Gamma_3^r \bar{\theta}_{st} \bar{\theta}_{uv} + \sum_{20} \Gamma_3^r \bar{\theta}_{stu} \bar{\theta}_v + J \sum_5 \bar{\theta}_{rstu} \bar{\theta}_v$$
$$+ \sum_5 \Gamma_2^r \bar{\theta}_{stuv} + M\bar{\theta}_r\bar{\theta}_s\bar{\theta}_t\bar{\theta}_u\bar{\theta}_v. \quad (2.23)$$

2.7.3 Contributing terms in the sampling moments

In Table 2.1 we give the terms in $\hat{\theta}$ which contribute to the moment to a given order in N^{-1}.

Table 2.1
Contributing terms in $\hat{\theta}$

Moment	Order	Terms
$E(\hat{\theta} - \theta)$	N^{-2}	Φ_3, Φ_4
	N^{-3}	Φ_4, Φ_5, Φ_6
$E\{(\hat{\theta} - \theta)^2\}$	N^{-3}	$\Phi_1\Phi_3, \Phi_2^2, \Phi_1\Phi_4, \Phi_2\Phi_3, \Phi_2\Phi_4, \Phi_3^2, \Phi_1\Phi_5$
$E\{(\hat{\theta} - \theta)^3\}$	N^{-3}	$\Phi_1^2\Phi_2, \Phi_1^2\Phi_3, \Phi_1\Phi_2^2, \Phi_1^2\Phi_4, \Phi_1\Phi_2\Phi_3, \Phi_2^3$
$E\{(\hat{\theta} - \theta)^4\}$	N^{-4}	$\Phi_1^3\Phi_2, \Phi_1^3\Phi_3, \Phi_1^2\Phi_2^2, \Phi_1^3\Phi_4, \Phi_1^2\Phi_2\Phi_3,$ $\Phi_1\Phi_2^3, \Phi_1^3\Phi_5, \Phi_1^2\Phi_2\Phi_4, \Phi_1^2\Phi_3^2, \Phi_1\Phi_2^2\Phi_3, \Phi_2^4$

2.7.4 Detailed consideration of the bias of $\hat{\theta}$

We denote the coefficient of N^{-s} in $E(\hat{\theta} - \theta)^r$ by $A_s^{(r)}$, so that

$$\mu_1'(\hat{\theta}) = \theta + \frac{A_1^{(1)}}{N} + \frac{A_2^{(1)}}{N} + \ldots, \tag{2.24}$$

where

$$A_1^{(1)} = \tfrac{1}{2} E_1 \Phi_2$$
$$= \tfrac{1}{2} E_1 (\epsilon_r \epsilon_s \bar{\theta}_{rs})$$
$$= E_1 (\epsilon_r \epsilon_s J \Gamma_1^r \Gamma_2^s + \epsilon_r \epsilon_s J \Gamma_2^r \Gamma_1^s + \epsilon_r \epsilon_s J \Gamma_1^r \Gamma_1^s)/(2I^3)$$

from (2.20b). Using (2.5b)

$$A_1^{(1)} = \{(p\Gamma_1 \Gamma_2) + \tfrac{1}{2}(p\Gamma_3)\}/I^2,$$

and since

$$(p\Gamma_3) + 2(p\Gamma_1 \Gamma_2) + \left(\Gamma_1 \frac{\partial}{\partial \theta} p\Gamma_1\right) = 0,$$

we also have

$$A_1^{(1)} = -E\left(\frac{1}{p^2} \frac{\partial p}{\partial \theta} \frac{\partial^2 p}{\partial \theta^2}\right) \bigg/ (2I^2),$$

which agrees with (2.10).

For $A_2^{(1)}$ we require $E_2 \Phi_3$ and $E_2 \Phi_4$. The former is $E \, \epsilon_r \epsilon_s \epsilon_t \bar{\theta}_{rst}$, for which we use (2.21b) in conjunction with (2.5a) and (2.5b), to obtain the contribution

$$3\{(p\Gamma_1^2 \Gamma_3) - IJ\}/I^3 + 6\{(p\Gamma_1 \Gamma_2^2) + 2I(p\Gamma_1 \Gamma_2)\}/I^3$$
$$+ 9J\{(p\Gamma_1^2 \Gamma_2) + I^2\}/I^4 + (KI + 3J^2)(p\Gamma_1^3)/I^5. \tag{2.25}$$

The contribution from Φ_4 consists of seven groups of terms as follows:

$$\lambda_1 = E_2 \sum_4 (\epsilon_r \epsilon_s \epsilon_t \epsilon_u \Gamma_4^r \bar{\theta}_s \bar{\theta}_t \bar{\theta}_u)/I,$$
$$\lambda_2 = K E_2 \sum_6 (\epsilon_r \epsilon_s \epsilon_t \epsilon_u \bar{\theta}_{rs} \bar{\theta}_t \bar{\theta}_u)/I,$$
$$\lambda_3 = J E_2 \sum_3 (\epsilon_r \epsilon_s \epsilon_t \epsilon_u \bar{\theta}_{rs} \bar{\theta}_{tu})/I,$$
$$\lambda_4 = E_2 \sum_{12} (\epsilon_r \epsilon_s \epsilon_t \epsilon_u \Gamma_3^r \bar{\theta}_s \bar{\theta}_{tu})/I,$$
$$\lambda_5 = J E_2 \sum_4 (\epsilon_r \epsilon_s \epsilon_t \epsilon_u \bar{\theta}_r \bar{\theta}_{stu})/I,$$
$$\lambda_6 = E_2 \sum_4 (\epsilon_r \epsilon_s \epsilon_t \epsilon_u \Gamma_2^r \bar{\theta}_{stu})/I,$$
$$\lambda_7 = L E_2 (\epsilon_r \epsilon_s \epsilon_t \epsilon_u \bar{\theta}_r \bar{\theta}_s \bar{\theta}_t \bar{\theta}_u)/I. \tag{2.26}$$

But from the last expression in (2.5a) and one derived from it by the operator $(i \, \partial/\partial h)(j \, \partial/\partial h)$, namely

$$E_2(h\epsilon)^2(i\epsilon)(j\epsilon) = 2\{(hip) - (hp)(ip)\}\{(hjp) - (hp)(jp)\}$$
$$+ \{(h^2p) - (hp)^2\}\{(ijp) - (ip)(jp)\}, \tag{2.27}$$

we have, using (2.19b), (2.20b) and (2.21b) in (2.26),

$$\lambda_1 = 12(p\Gamma_1\Gamma_4)/I^3,$$

$$\lambda_2 = 18K\{2(p\Gamma_1\Gamma_2) + J\}/I^4,$$

$$\lambda_3 = 3J\{8(p\Gamma_1\Gamma_2)^2 + 4I(p\Gamma_2^2) + 12J(p\Gamma_1\Gamma_2) + 3J^2 - 4I^3\}/I^5,$$

$$\lambda_4 = 12\{4(p\Gamma_1\Gamma_2)(p\Gamma_1\Gamma_3) + 2I(p\Gamma_2\Gamma_3) + 2I^2J + 3J(p\Gamma_1\Gamma_3)\}/I^4,$$

$$\lambda_5 = 4J\{9I(p\Gamma_1\Gamma_3) + 27J(p\Gamma_1\Gamma_2) + 9J^2 + 3IK + 12(p\Gamma_1\Gamma_2)^2$$
$$+ 6I(p\Gamma_2^2) - 6I^3\}/I^5,$$

$$\lambda_6 = 4\{6I(p\Gamma_1\Gamma_2)(p\Gamma_1\Gamma_3) + 3I^2(p\Gamma_2\Gamma_3) + 18J(p\Gamma_1\Gamma_2)^2$$
$$+ 9IJ(p\Gamma_2^2) - 6I^3J + 18I(p\Gamma_1\Gamma_2)(p\Gamma_2^2) + (9J^2 + 3IK - 18I^3)$$
$$\times (p\Gamma_1\Gamma_2)\}/I^5,$$

$$\lambda_7 = 3L/I^3. \tag{2.28}$$

Hence, adding the contributions (2.25) and (2.28) leads to the N^{-2} coefficient in the bias.

2.7.5 Sufficient detail has now been given to demonstrate the evaluation of the moments by this indirect approach. We now list the first four moments, using the abbreviated notation for products of logarithmic derivatives

$$(p\Gamma_1^{r_1}\Gamma_2^{r_2}\ldots\Gamma_k^{r_k}) = [1^{r_1}\,2^{r_2}\,\ldots k^{r_k}]. \tag{2.29}$$

2.7.6 *First four moments using logarithmic derivatives*

Mean

$$\mu_1'(\hat{\theta}) \sim \theta + \sum_1^3 A_s^{(1)}/N^s, \tag{2.30a}$$

where

$$A_1^{(1)} = \{[12] + \tfrac{1}{2}[3]\}/I^2,$$
$$A_2^{(1)} = \sum_{s=2}^{5} a_s^{(1)}/I^s,$$

where

$$a_2^{(1)} = -\{\tfrac{1}{2}[3] + [12]\},$$

$$8a_3^{(1)} = 4[1^23] + 8[12^2] + 4[14] + 12[23] + [5],$$

$$12a_4^{(1)} = 2[4][1^3] + 18[3][1^22] + 15[3][4] + 36[3][2^2]$$
$$+ 36[3][13] + 36[12][13] + 36[12][2^2] + 24[4][12],$$

$$8a_5^{(1)} = 4[3]^2[1^3] + 15[3]^3 + 48[3][12]^2 + 60[3]^2[12],$$

$$A_3^{(1)} = \sum_{s=2}^{8} b_s^{(1)}/I^s,$$

where

$$b_2^{(1)} = [12] + \tfrac{1}{2}[3],$$

$$8b_3^{(1)} = -\{12[14] + 12[1^2 3] + 36[23] + 24[12^2] + 3[5]\},$$

$$\begin{aligned}48b_4^{(1)} &= 8[1^3 4] + 72[1^2 23] + 48[12^3] + 12[1^2 5] + 96[124] \\ &\quad + 144[2^2 3] + 72[13^2] + 6[16] + 30[25] + 60[34] \\ &\quad - 432[13][12] - 432[12][2^2] - 24[4][1^3] - 288[4][12] \\ &\quad + [7] - [3]\{180[4] + 216[1^2 2] + 432[13] + 432[2^2]\},\end{aligned}$$

$$\begin{aligned}48b_5^{(1)} &= 3[3]\{16[1^3 3] + 48[1^2 2^2] - 288[12]^2 + 40[1^2 4] \\ &\quad + 240[123] + 80[2^3] + 30[15] + 120[24] - 24[3][1^3] \\ &\quad - 360[3][12] - 90[3]^2 + 90[3^2] + 7[6]\} + 8[4]\{4[1^3 2] \\ &\quad + 15[1^2 3] + 30[12^2] + 15[14] + 45[23]\} + [5]\{2[1^4] \\ &\quad + 60[1^2 2] + 90[13] + 90[2^2] + 35[4]\} + 4[1^3]\{2[15] \\ &\quad + 8[24] + 6[3^2] + [6]\} + 12[12]\{8[1^2 4] + 48[123] \\ &\quad + 16[2^3] + 10[15] + 40[24] + 30[3^2] + 3[6]\} \\ &\quad + 96[1^2 2][14] + 288[12^2][13] + 288[1^2 2][23] \\ &\quad + 144[1^2 3][2^2] + 144[1^2 3][13] + 288[12^2][2^2] \\ &\quad + 240[13][14] + 240[14][2^2] + 720[23][13] \\ &\quad + 720[2^2][23],\end{aligned}$$

$$\begin{aligned}24b_6^{(1)} &= [3]\{10[4][1^4] + 360[1^2 2][2^2] + 105[3][5] + 140[4]^2 \\ &\quad + 360[4][1^2 2] + 540[3][12^2] + 60[3][1^3 2] + 40[1^3][14] \\ &\quad + 360[1^2 2][13] + 360[1^2 3][12] + 120[1^3][23] \\ &\quad + 270[3][1^2 3] + 30[5][1^3] + 540[13]^2 + 1080[13][2^2] \\ &\quad + 540[2^2]^2 + 315[3][14] + 630[4][13] + 945[3][23]\} \\ &\quad + 2[4]\{20[1^3][13] + 120[1^2 2][12] + 20[1^3][2^2] \\ &\quad + 10[4][1^3] + 315[3][2^2]\} + [12]\{20[5][1^3] \\ &\quad + 240[14][12] + 360[13]^2 + 720[23][12] + 720[2^2][13]\end{aligned}$$

$$\begin{aligned}
&+ 360[2^2]^2 + 720[3][14] + 720[4][13] + 2160[3][23] \\
&+ 720[4][2^2] + 180[5][12] + 315[5][3] + 210[4]^2 \\
&+ 720[3][12^2]\},
\end{aligned}$$

$$\begin{aligned}
8b_7^{(1)} = \ & [3]^2\{5[3][1^4] + 60[1^3][13] + 70[4][1^3] + 210[3][1^22] \\
&+ 360[1^22][12] + 60[1^3][2^2] + 1260[12][13] \\
&+ 1260[2^2][12] + 420[3][13] + 420[3][2^2] + 480[4][12] \\
&+ 210[4][3]\} + 8[3]\{10[4][1^3][12] + 90[13][12]^2 \\
&+ 90[2^2][12]^2 + 105[4][12]^2\} + 160[4][12]^3,
\end{aligned}$$

$$\begin{aligned}
16b_8^{(1)} = \ & [3]^2\{140[3]^2[1^3] + 280[3][1^3][12] + 1680[12]^3 \\
&+ 3360[3][12]^2 + 1890[3]^2[12] + 315[3]^3\}.
\end{aligned}$$

Variance

$$\mu_2(\hat{\theta}) \sim \sum_{s=1}^{3} B_s^{(2)}/N^s, \qquad (2.30b)$$

where

$$\begin{aligned}
B_1^{(2)} = \ & 1/I, \\
B_2^{(2)} = \ & -1/I + (3[2^2] + 3[13] + 2[1^22] + [4])/I^3 \\
&+ (5[12]^2 + 11[3][12] + 7[3]^2/2 + [3][1^3])/I^4, \\
B_3^{(2)} = \ & \sum_{s=1}^{7} c_s^{(2)}/I^s,
\end{aligned}$$

where

$$\begin{aligned}
c_1^{(2)} = \ & 1, \\
c_2^{(2)} = \ & 0, \\
c_3^{(2)} = \ & -3[4] - 9[13] - 6[1^22] - 9[2^2], \\
4c_4^{(2)} = \ & 4[1^33] + 12[1^22^2] - 64[12]^2 - 12[3][1^3] - 136[3][12] \\
&- 43[3]^2 + 16[2^3] + 48[123] + 8[1^24] + 20[24] + 15[3^2] \\
&+ 5[15] + [6], \\
12c_5^{(2)} = \ & 4[4][1^4] + 48[3][1^32] + 144[1^22][13] + 132[1^23][12] \\
&+ 48[1^3][23] + 144[1^22][2^2] + 264[12^2][12] \\
&+ 348[3][12^2] + 174[3][1^23] + 120[4][1^22] \\
&+ 16[1^3][14] + 10[5][1^3] + 228[12][14] + 684[23][12]
\end{aligned}$$

$$\begin{aligned}
&+ 360[13][2^2] + 522[3][23] + 180[4][2^2] + 180[2^2]^2\\
&+ 87[5][12] + 174[3][14] + 180[13]^2 + 180[4][13]\\
&+ 51[5][3] + 35[4]^2,
\end{aligned}$$

$$\begin{aligned}
12c_6^{(2)} &= 15[3]^2[1^4] + 684[3][1^22][12] + 120[3][1^3][2^2]\\
&+ 522[3]^2[1^22] + 120[3][1^3][13] + 76[4][1^3][12]\\
&+ 118[4][3][1^3] + 648[13][12]^2 + 2052[3][2^2][12]\\
&+ 2052[12][13][3] + 672[4][12]^2 + 909[3]^2[2^2]\\
&+ 648[2^2][12]^2 + 1206[4][3][12] + 909[3]^2[13]\\
&+ 405[4][3]^2,
\end{aligned}$$

$$\begin{aligned}
4c_7^{(2)} &= 116[3]^2[1^3][12] + 68[3]^3[1^3] + 432[3][12]^3\\
&+ 1176[3]^2[12]^2 + 795[3]^3[12] + 150[3]^4.
\end{aligned}$$

Third central moment

$$\mu_3(\hat{\theta}) \sim \sum_{s=2}^{3} B_s^{(3)}/N^s, \tag{2.30c}$$

where

$$B_2^{(3)} = \{[1^3] + 6[12] + 3[3]\}/I^3,$$
$$B_3^{(3)} = \sum_{s=3}^{6} b_s^{(3)}/I^s,$$

where

$$\begin{aligned}
2b_3^{(3)} &= -\{6[1^3] + 42[12] + 21[3]\},\\
2b_4^{(3)} &= 6[1^32] + 15[1^23] + 30[12^2] + 12[14] + 36[23] + 3[5],\\
2b_5^{(3)} &= 3[3][1^4] + 12[1^3][13] + 60[1^22][12] + 12[1^3][2^2]\\
&+ 75[3][1^22] + 9[4][1^3] + 144[13][12] + 108[3][13]\\
&+ 72[4][12] + 42[3][4] + 144[12][2^2] + 108[3][2^2],\\
2b_6^{(3)} &= 54[3][1^3][12] + 39[3]^2[1^3] + 88[12]^3 + 420[3][12]^2\\
&+ 372[3]^2[12] + 83[3]^3.
\end{aligned}$$

Fourth central moment

$$\mu_4(\hat{\theta}) \sim \sum_{s=2}^{4} B_s^{(4)}/N^s, \tag{2.30d}$$

where

$B_2^{(4)} = 3/I^2,$

$B_3^{(4)} = \sum_{s=2}^{5} b_s^{(4)}/I^s,$

where

$b_2^{(4)} = -9,$

$b_3^{(4)} = 0,$

$b_4^{(4)} = [1^4] + 24[1^22] + 30[13] + 30[2^2] + 10[4],$

$b_5^{(4)} = 45[3]^2 + 12[1^3][12] + 18[3][1^3] + 90[12]^2 + 150[3][12],$

$B_4^{(4)} = \sum_{s=2}^{8} c_s^{(4)}/I^s,$

where

$c_2^{(4)} = 21,$

$c_3^{(4)} = 0,$

$c_4^{(4)} = -6[1^4] - 144[1^22] - 180[13] - 180[2^2] - 60[4],$

$2c_5^{(4)} = 8[1^42] - 160[1^3][12] + 40[1^33] - 1188[12]^2 + 120[1^22^2]$
$\qquad + 60[1^24] + 360[123] - 224[3][1^3] - 1908[3][12]$
$\qquad - 567[3]^2 + 120[2^3] + 35[15] + 140[24] + 105[3^2] + 7[6],$

$4c_6^{(4)} = 8[3][1^5] + 72[1^3][1^23] + 40[1^4][13] + 40[4][1^4]$
$\qquad + 456[3][1^32] + 240[1^22]^2 + 144[1^3][12^2] + 40[1^4][2^2]$
$\qquad + 272[1^32][12] + 152[1^3][14] + 840[4][1^22] + 68[5][1^3]$
$\qquad + 2328[3][12^2] + 456[1^3][23] + 1248[1^23][12]$
$\qquad + 1440[1^22][13] + 1164[3][1^23] + 2496[12^2][12]$
$\qquad + 1440[1^22][2^2] + 1512[12][14] + 1036[3][14]$
$\qquad + 1120[4][2^2] + 210[4]^2 + 518[5][12] + 1120[4][13]$
$\qquad + 3108[3][23] + 4536[23][12] + 2520[13][2^2]$
$\qquad + 1260[2^2]^2 + 1260[13]^2 + 294[3][5],$

$2c_7^{(4)} = 12[4][1^3]^2 + 99[3]^2[1^4] + 228[3][1^3][1^22]$
$\qquad + 108[3][1^4][12] + 792[3][1^3][2^2] +$
$\qquad + 4488[3][1^22][12] + 408[1^3][2^2][12] + 1176[1^22][12]^2$
$\qquad + 496[4][1^3][12] + 729[3][1^3][13] + 408[1^3][12][13]$

$$\begin{aligned}
&+ 2310[3]^2[1^22] + 526[3][4][1^3] + 3447[3]^2[2^2] \\
&+ 2916[4][12]^2 + 1443[3]^2[4] + 9000[3][12][2^2] \\
&+ 4554[3][4][12] + 3447[3]^2[13] + 9000[3][12][13] \\
&+ 4212[12]^2[13] + 4212[2^2][12]^2,
\end{aligned}$$

$$\begin{aligned}
4c_8^{(4)} =\ & 132[3]^2[1^3]^2 + 5976[3]^2[1^3][12] + 2292[3]^3[1^3] \\
& + 2808[1^3][12]^2[3] + 2532[12]^4 + 38808[3]^2[12]^2 \\
& + 3843[3]^4 + 21912[3][12]^3 + 22086[3]^3[12].
\end{aligned}$$

2.7.7 Skewness and kurtosis

Measure of skewness

$$\gamma_1(\hat{\theta}) = \mu_3/\mu_2^{3/2}$$
$$\sim \frac{G_1}{\sqrt{(NI)}} + \frac{G_2}{(NI)^{3/2}}, \qquad (2.31a)$$

where

$$G_1 = \{[1^3] + 6[12] + 3[3]\}/I,$$
$$G_2 = \sum_{s=1}^{4} e_s/I^{s-1},$$

where

$$2e_1 = -\{3[1^3] + 24[12] + 12[3]\},$$
$$2e_2 = 6[1^32] + 15[1^23] + 30[12^2] + 12[14] + 36[23] + 3[5],$$
$$\begin{aligned}
2e_3 =\ & 3[3][1^4] + 3[1^3][13] + 24[1^22][12] + 3[1^3][2^2] \\
& + 6[4][1^3] + 90[13][12] + 81[3][13] + 54[4][12] \\
& + 33[3][4] + 90[12][2^2] + 81[3][2^2] - 6[1^3][1^22],
\end{aligned}$$
$$\begin{aligned}
4e_4 =\ & 6[3][1^3][12] - 4[12]^3 + 354[3][12]^2 + 420[3]^2[12] \\
& + 103[3]^3 - 30[1^3][12]^2 - 6[3][1^3]^2 + 39[3]^2[1^3].
\end{aligned}$$

Measure of kurtosis

$$\gamma_2(\hat{\theta}) = \mu_4/\mu_2^2 - 3$$
$$\sim \frac{1}{N}\sum_{s=0}^{3} f_s I^{-s} + \frac{1}{N^2}\sum_{s=0}^{6} g_s I^{-s}, \qquad (2.31b)$$

where

$$f_0 = -3,$$
$$f_1 = 0,$$

$$f_2 = [1^4] + 12[1^2 2] + 12[13] + 12[2^2] + 4[4],$$

$$f_3 = 24[3]^2 + 12[1^3][12] + 12[3][1^3] + 60[12]^2 + 84[3][12],$$

$$g_0 = 6,$$

$$g_1 = 0,$$

$$g_2 = -4[1^4] - 60[1^2 2] - 66[13] - 66[2^2] - 22[4],$$

$$\begin{aligned}g_3 =\ & 4[1^4 2] - 56[1^3][12] + 14[1^3 3] - 318[12]^2 + 42[1^2 2^2] \\ & + 18[1^2 4] + 108[123] - 58[3][1^3] - 450[3][12] - 129[3]^2 \\ & + 36[2^3] + 10[15] + 40[24] + 30[3^2] + 2[6],\end{aligned}$$

$$\begin{aligned}g_4 =\ & 2[3][1^5] + 18[1^3][1^2 3] + 4[1^4][13] + 6[4][1^4] \\ & + 90[3][1^3 2] + 36[1^3][12^2] + 4[1^4][2^2] + 68[1^3 2][12] \\ & + 30[1^3][14] + 98[4][1^2 2] + 12[5][1^3] + 408[3][12^2] \\ & + 90[1^3][23] + 246[1^2 3][12] + 132[1^2 2][13] + 204[3][1^2 3] \\ & + 492[12^2][12] + 132[1^2 2][2^2] + 264[12][14] \\ & + 172[3][14] + 124[4][2^2] + 24[4]^2 + 124[4][13] \\ & + 516[3][23] + 792[23][12] + 252[13][2^2] + 126[2^2]^2 \\ & + 126[13]^2 + 48[3][5] - 4[1^4][1^2 2],\end{aligned}$$

$$\begin{aligned}g_5 =\ & 6[4][1^3]^2 + 35[3]^2[1^4] + 30[3][1^3][1^2 2] + 32[3][1^4][12] \\ & + 222[3][1^3][2^2] + 1170[3][1^2 2][12] + 132[1^3][2^2][12] \\ & + 168[1^2 2][12]^2 + 186[4][1^3][12] + 222[3][1^3][13] \\ & + 132[1^3][12][13] + 672[3]^2[1^2 2] + 166[3][4][1^3] \\ & + 978[3]^2[2^2] + 932[4][12]^2 + 422[3]^2[4] \\ & + 2508[3][12][2^2] + 1352[3][4][12] + 978[3]^2[13] \\ & + 2508[3][12][13] + 1212[12]^2[13] + 1212[2^2][12]^2 \\ & - 48[1^2 2][1^3][12] - 10[1^4][12]^2 - 2[3][1^3][1^4],\end{aligned}$$

$$\begin{aligned}g_6 =\ & 6[3]^2[1^3]^2 + 738[3]^2[1^3][12] + 318[3]^3[1^3] \\ & + 168[1^3][12]^2[3] - 42[12]^4 + 4962[3]^2[12]^2 + 531[3]^4 \\ & + 2340[12]^3[3] + 2982[3]^3[12] - 120[1^3][12]^3 \\ & - 24[3][1^3]^2[12].\end{aligned}$$

2.8 Moments using derivatives of the probability function

2.8.1 There may be structures for which derivatives of $p_r(\theta)$ may be simpler than derivatives of $\log p_r(\theta)$. The expressions given in (2.30) have been converted to this notation. We now define

$$E\left(\left(\frac{1}{p}\frac{\partial p}{\partial \theta}\right)^a \left(\frac{1}{p}\frac{\partial^2 p}{\partial \theta^2}\right)^b \left(\frac{1}{p}\frac{\partial^3 p}{\partial \theta^3}\right)^c \ldots\right) = (1^a 2^b 3^c \ldots). \quad (2.32)$$

For example, in the discrete case

$$(1^2) = \sum \frac{1}{p}\left(\frac{\partial p}{\partial \theta}\right)^2, \quad (1^2 2) = \sum \frac{1}{p^2}\left(\frac{\partial p}{\partial \theta}\right)^2 \frac{\partial^2 p}{\partial \theta^2},$$

and

$$(p\Gamma_1\Gamma_2) = (12) - (1^3),$$
$$(p\Gamma_1^2\Gamma_3) = (1^2 3) - 3(1^3 2) + 2(1^5),$$

and so on. In this notation, we find for the first four moments of $\hat{\theta}$:

Mean

$$\mu_1'(\hat{\theta}) = \theta - \frac{(12)}{2NI^2} + \sum_{s=2}^{5} \frac{a_s^{(1)} I^{-s}}{N^2} + \sum_{s=2}^{8} \frac{b_s^{(1)} I^{-s}}{N^3} + \ldots \quad (2.33a)$$

where

$$2a_2^{(1)} = (12),$$
$$8a_3^{(1)} = 20(1^3 2) + 2(23) - 4(1^2 3) - 8(1^5) - 10(12^2) - (14),$$
$$12a_4^{(1)} = 12(12)(13) + 54(12)(1^2 2) + 4(1^3)(13) + 24(1^3)(1^4)$$
$$\quad + 12(1^3)(2^2) - 36(12)(1^4) - 9(12)(2^2) - 48(1^3)(1^2 2),$$
$$8a_5^{(1)} = \{2(1^3) - 3(12)\}\{12(1^3)(12) - 4(1^3)^2 + 3(12)^2\},$$

and

$$b_2^{(1)} = -\tfrac{1}{2}(12),$$
$$8b_3^{(1)} = 30(12^2) + 3(14) - 6(23) + 24(1^5) - 60(1^3 2) + 12(1^2 3),$$
$$48b_4^{(1)} = -126(2^2 3) + 25(34) + 9(25) + 648(1^2 23) - 12(1^2 5)$$
$$\quad + 318(12^3) - 88(13^2) - 114(124) - (16) - 240(1^7)$$
$$\quad + 960(1^5 2) - 1104(1^3 2^2) - 368(1^4 3) + 92(1^3 4)$$
$$\quad + 108(12)(2^2) - 144(1^3)(2^2) - 144(12)(13) - 48(1^3)(13)$$
$$\quad + 432(1^4)(12) - 648(1^2 2)(12) - 288(1^4)(1^3)$$
$$\quad + 576(1^2 2)(1^3),$$
$$48b_5^{(1)} = 558(12^2)(2^2) - 45(14)(2^2) - 210(23)(2^2) + 312(1^5)(2^2)$$
$$\quad - 924(1^3 2)(2^2) + 324(1^2 3)(2^2) - 195(12)(24) + 122(1^3)(24)$$

$$\begin{aligned}
&- 180(12)(3^2) + 104(1^3)(3^2) + 12(12)(15) + 8(1^3)(15) \\
&+ 4104(12)(1^4 2) - 1776(1^3)(1^4 2) - 1464(12)(1^3 3) \\
&+ 720(1^3)(1^3 3) - 3564(12)(1^2 2^2) + 1656(1^3)(1^2 2^2) \\
&+ 462(12)(2^3) - 228(1^3)(2^3) + 1776(12)(123) \\
&- 928(1^3)(123) + 276(12)(1^2 4) - 168(1^3)(1^2 4) \\
&+ 708(13)(12^2) + 636(1^4)(12^2) - 1692(1^2 2)(12^2) \\
&+ 10(13)(14) - 106(1^4)(14) + 186(1^2 2)(14) - 220(13)(23) \\
&- 308(1^4)(23) + 708(1^2 2)(23) + 464(1^5)(13) - 1256(1^3 2)(13) \\
&+ 344(1^2 3)(13) - 1200(12)(1^6) + 480(1^3)(16) \\
&- 648(1^3)^2(12) + 240(1^4)(1^5) - 816(1^2 2)(1^5) \\
&- 888(1^4)(1^3 2) + 2616(1^2 2)(1^3 2) + 424(1^4)(1^2 3) \\
&- 1032(1^2 2)(1^2 3) + 540(1^3)(12)^2 + 162(12)^3 + 144(1^3)^3,
\end{aligned}$$

$$\begin{aligned}
24 b_6^{(1)} =\ & 630(12)(13)(2^2) - 460(1^3)(13)(2^2) + 1080(1^4)(12)(2^2) \\
&- 2430(1^2 2)(12)(2^2) - 480(1^4)(1^3)(2^2) + 1260(1^2 2)(1^3)(2^2) \\
&+ 3030(1^3)(12)(12^2) - 2880(12)^2(12^2) - 660(1^3)^2(12^2) \\
&- 205(1^3)(12)(14) + 15(12)^2(14) + 90(1^3)^2(14) \\
&- 1190(1^3)(12)(23) + 945(12)^2(23) + 320(1^3)^2(23) \\
&+ 1260(1^4)(12)(13) - 2400(1^2 2)(12)(13) - 680(1^4)(1^3)(13) \\
&+ 1520(1^2 2)(1^3)(13) + 1560(1^5)(1^3)(12) - 4800(1^3 2)(1^3)(12) \\
&+ 1720(1^2 3)(1^3)(12) - 3060(1^2 2)(1^4)(12) + 540(1^4)^2(12) \\
&+ 3780(1^2 2)^2(12) + 1080(1^2 2)(1^4)(1^3) - 120(1^4)^2(1^3) \\
&- 1680(1^2 2)^2(1^3) + 360(12)(2^2)^2 - 210(1^3)(2^2)^2 \\
&+ 60(12)(13)^2 - 160(1^3)(13)^2 - 1800(1^5)(12)^2 \\
&+ 4995(1^3 2)(12)^2 - 1455(1^2 3)(12)^2 - 240(1^5)(1^3)^2 \\
&+ 900(1^3 2)(1^3)^2 - 420(1^2 3)(1^3)^2,
\end{aligned}$$

$$\begin{aligned}
8 b_7^{(1)} =\ & - 840(1^3)^2(12)(2^2) + 1530(1^3)(12)^2(2^2) - 750(12)^3(2^2) \\
&+ 120(1^3)^3(2^2) - 880(1^3)^2(12)(13) + 1200(1^3)(12)^2(13) \\
&- 280(12)^3(13) + 160(1^3)^3(13) - 780(1^3)^2(1^4)(12) \\
&+ 2280(1^3)^2(12)(1^2 2) - 2070(1^3)(1^4)(12)^2
\end{aligned}$$

$$\begin{aligned}
& \quad - 4860(1^22)(1^3)(12)^2 - 1365(1^4)(12)^3 + 2730(1^22)(12)^3 \\
& \quad + 40(1^4)(1^3)^3 - 240(1^22)(1^3)^3, \\
16b_8^{(1)} & = 560(1^3)^4(12) - 3360(1^3)^3(12)^2 + 6720(1^3)^2(12)^3 \\
& \quad - 5040(1^3)(12)^4 + 945(12)^5.
\end{aligned}$$

Variance

$$\mu_2(\hat{\theta}) = \frac{1}{NI} + \frac{1}{N^2} \sum_1^4 b_s I^{-s} + \frac{1}{N^3} \sum_{s=1}^7 c_s I^{-s} + \ldots \quad (2.33b)$$

where

$$\begin{aligned}
b_1 & = -1, \\
b_2 & = 0, \\
b_3 & = (1^4) - (1^22) - (13), \\
2b_4 & = 7(12)^2 - 2(1^3)^2,
\end{aligned}$$

and

$$\begin{aligned}
c_1 & = 1, \\
c_2 & = 0, \\
c_3 & = -3(1^4) + 3(1^22) + 3(13), \\
4c_4 & = 5(24) + 111(1^22^2) - 14(2^3) - 132(1^42) - 7(1^24) + 44(1^33) \\
& \quad + 40(1^6) - (15) - 52(123) + 5(3^2) + 12(1^3)^2 - 43(12)^2, \\
12c_5 & = 6(12)\{6(14) + 58(1^23) - 232(1^32) + 128(12^2) + 88(1^5) \\
& \quad - 37(23)\} + (1^3)\{11(14) - 236(1^23) + 756(1^32) - 450(12^2) \\
& \quad - 264(1^5) + 158(23)\} + 20(13)^2 - 60(13)(2^2) - 160(13)(1^4) \\
& \quad + 264(1^22)(13) + 324(1^22)(2^2) + 480(1^22)(1^4) - 45(2^2)^2 \\
& \quad - 156(1^4)(2^2) - 96(1^4)^2 - 540(1^22)^2, \\
12c_6 & = 360(1^4)(1^3)^2 - 123(13)(12)^2 - 496(13)(12)(1^3) \\
& \quad + 260(13)(1^3)^2 - 2331(1^22)(12)^2 + 3252(1^22)(12)(1^3) \\
& \quad - 900(1^22)(1^3)^2 + 576(12)^2(2^2) - 948(12)(1^3)(2^2) \\
& \quad + 300(1^3)^2(2^2) + 1314(1^4)(12)^2 - 1536(1^4)(12)(1^3), \\
4c_7 & = \{2(1^3) - 3(12)\}\{9(12)^3 - 219(12)^2(1^3) + 148(12)(1^3)^2 \\
& \quad - 20(1^3)^3\}.
\end{aligned}$$

Third central moment

$$\mu_3(\hat{\theta}) = \frac{(1^3) - 3(12)}{N^2 I^3} + \sum_{s=3}^{6} \frac{d_s I^{-s}}{N^3} + \ldots, \quad (2.33c)$$

where

$$2d_3 = 21(12) - 6(1^3),$$
$$2d_4 = 45(1^3 2) - 18(1^5) + 6(23) - 9(1^2 3) - 24(12^2) - 3(14),$$
$$2d_5 = (12)\{36(13) + 87(1^2 2) - 60(1^4) - 18(2^2)\}$$
$$\quad + (1^3)\{42(1^4) - 78(1^2 2) + 21(2^2)\},$$
$$2d_6 = -65(12)^3 - 45(12)^2(1^3) + 90(12)(1^3)^2 - 24(1^3)^3.$$

Fourth central moment

$$\mu_4(\hat{\theta}) = \frac{3}{N^2 I^2} + \sum_{s=2}^{5} \frac{e_s I^{-s}}{N^3} + \sum_{s=2}^{8} \frac{f_s I^{-s}}{N^4} + \ldots, \quad (2.33d)$$

where

$$e_2 = -9,$$
$$e_3 = 0,$$
$$e_4 = 7(1^4) - 6(1^2 2) - 10(13),$$
$$e_5 = -6(1^3)^2 - 12(1^3)(12) + 45(12)^2,$$
$$f_2 = 21,$$
$$f_3 = 0,$$
$$f_4 = -42(1^4) + 60(13) + 36(1^2 2),$$
$$2f_5 = 252(1^6) + 72(1^3)^2 - 832(1^4 2) + 152(1^3)(12) + 280(1^3 3) - 7(15)$$
$$\quad - 45(1^2 4) - 340(123) + 35(3^2) + 705(1^2 2^2) - 90(2^3) + 35(24)$$
$$\quad - 567(12)^2,$$
$$2f_6 = (1^3)\{-624(1^5) + 1760(1^3 2) + 11(14) - 536(1^2 3) + 378(23)$$
$$\quad - 1050(12^2)\} + 2(1^4)\{520(1^2 2) - 170(2^2) - 180(13) - 105(1^4)\}$$
$$\quad + 2(12)\{56(14) + 412(1^2 3) - 287(23) - 1660(1^3 2) + 936(12^2)$$
$$\quad + 628(1^5)\} - 105(2^2)^2 + 70(13)^2 - 1170(1^2 2)^2 + 580(13)(1^2 2)$$
$$\quad - 140(13)(2^2) + 720(2^2)(1^2 2),$$
$$2f_7 = 2(1^3)^2\{462(1^4) + 292(13) - 1104(1^2 2) + 366(2^2)\}$$
$$\quad + 2(1^3)\{3972(12)(1^2 2) - 1920(12)(1^4) - 1188(12)(2^2)$$

$$-458(12)(13)\} + (12)^2 \{3324(1^4) - 5781(1^2 2) + 1512(2^2)$$
$$- 729(13)\},$$
$$4f_8 = -684(1^3)^4 + 5640(1^3)^3(12) - 13320(1^3)^2(12)^2 + 9018(1^3)(12)^3$$
$$+ 1029(12)^4.$$

2.8.2 Skewness and kurtosis

Measure of skewness

$$\gamma_1(\hat{\theta}) \sim \frac{G_1}{\sqrt{(NI)}} + \frac{G_2}{(NI)^{3/2}}, \tag{2.34a}$$

where

$$G_1 = \{(1^3) - 3(12)\}/I,$$
$$G_2 = \sum_{s=1}^{4} e_s/I^{s-1},$$

where

$$2e_1 = 12(12) - 3(1^3),$$
$$2e_2 = 45(1^3 2) - 18(1^5) + 6(23) - 9(1^2 3) - 24(12^2) - 3(14),$$
$$2e_3 = (12)\{27(13) + 78(1^2 2) - 51(1^4) - 18(2^2)\}$$
$$\qquad + (1^3)\{39(1^4) - 75(1^2 2) + 21(2^2) + 3(13)\},$$
$$4e_4 = -42(1^3)^3 + 162(1^3)^2(12) - 111(1^3)(12)^2 - 67(12)^3.$$

Measure of kurtosis

$$\gamma_2(\hat{\theta}) \sim \frac{1}{N} \sum_{s=0}^{3} f_s I^{-s} + \frac{1}{N^2} \sum_{s=0}^{6} g_s I^{-s}, \tag{2.34b}$$

where

$$f_0 = -3,$$
$$f_1 = 0,$$
$$f_2 = (1^4) - 4(13),$$
$$f_3 = 24(12)^2 - 12(1^3)(12),$$

and

$$g_0 = 6,$$
$$g_1 = 0,$$
$$g_2 = -10(1^4) + 22(13) + 6(1^2 2),$$
$$g_3 = 66(1^6) + 6(1^3)^2 - 218(1^4 2) + 52(1^3)(12) - 129(12)^2 + 10(24)$$

$$+ 186(1^2 2^2) - 24(2^3) + 74(1^3 3) - 2(15) - 12(1^2 4) + 10(3^2)$$
$$- 92(123),$$

$$g_4 = (1^3)\{502(1^3 2) - 180(1^5) - 150(1^2 3) + 110(23) - 300(12^2)\}$$
$$+ (1^4)\{-62(1^4) - 84(13) - 92(2^2) + 288(1^2 2)\} + (12)\{38(14)$$
$$+ 238(1^2 3) - 176(23) - 964(1^3 2) + 552(12^2) + 364(1^5)\}$$
$$- 30(2^2)^2 + 14(13)^2 - 318(1^2 2)^2 + 144(13)(1^2 2) - 40(13)(2^2)$$
$$+ 198(1^2 2)(2^2),$$

$$g_5 = (1^3)^2 \{290(1^4) + 148(13) - 660(1^2 2) + 216(2^2)\}$$
$$+ (1^3)\{-1128(12)(1^4) + 2322(12)(1^2 2) - 714(12)(2^2)$$
$$- 234(12)(13)\} + (12)^2 \{929(1^4) - 1656(1^2 2) + 468(2^2)$$
$$- 206(13)\},$$

$$g_6 = -114(1^3)^4 + 852(1^3)^3(12) - 1938(1^3)^2(12)^2 + 1326(1^3)(12)^3$$
$$+ 93(12)^4.$$

2.9 Illustrations

2.9.1 Probability function linear in θ

Clearly, $\partial^s p_r / \partial \theta^s = 0, s > 1$, so that the moments of $\hat{\theta}$ are functions of the first derivatives only, and the notation of **2.8** is appropriate. Define therefore

$$A_s = E\left(\frac{1}{p} \frac{\partial p}{\partial \theta}\right)^{s+1}$$
$$= (1^{s+1}).$$

Noting that $(1^a 2^b 3^c \ldots) = 0$ if b or c or $\ldots \neq 0$, and using (2.33),

$$\mu_1'(\hat{\theta}) \sim \theta + A_1^{(1)}/N + A_2^{(1)}/N^2 + A_3^{(1)}/N^3, \qquad (2.35a)$$

where

$$A_1^{(1)} = 0,$$
$$A_1^5 A_2^{(1)} = -(A_2^3 - 2A_1 A_2 A_3 + A_1^2 A_4),$$
$$A_1^7 A_3^{(1)} = 3A_4 A_1^4 - (5A_6 + 6A_2 A_3)A_1^3 + (10A_2 A_5 + 3A_3^2 + 5A_3 A_4)A_1^2$$
$$- (5A_3^2 A_2 + 10A_4 A_2^2)A_1 + 5A_3 A_2^3,$$
$$\mu_2(\hat{\theta}) \sim 1/(NA_1) + (A_3 A_1 - A_2^2 - A_1^3)/(N^2 A_1^4) + (A_1^6 - 3A_3 A_1^4$$
$$+ 10A_5 A_1^3 + 3A_2^2 A_1^3 - 8A_3^2 A_1^2 - 22A_4 A_2 A_1^2 + 30A_3 A_2^2 A_1$$
$$- 10A_2^4)/(N^3 A_1^7), \qquad (2.35b)$$

$$\mu_3(\hat{\theta}) \sim A_2/(N^2 A_1^3) - (12A_2^3 - 21A_3 A_2 A_1 + 9A_4 A_1^2 + 3A_2 A_1^3)/(N^3 A_1^6),$$
(2.35c)

$$\mu_4(\theta) \sim 3/(N^2 A_1^2) - (6A_2^2 - 7A_3 A_1 + 9A_1^3)/(N^3 A_1^5) + (21A_1^6 - 42A_3 A_1^4$$
$$+ 126A_5 A_1^3 + 36A_2^2 A_1^3 - 312A_4 A_2 A_1^2 - 105A_3^2 A_1^2$$
$$+ 462A_3 A_2^2 A_1 - 171A_2^4)/(N^4 A_1^8).$$
(2.35d)

The first two terms in (2.35b), the first term in (2.35c), and the first two terms in (2.35d) agree with those given by Haldane and Smith (1956); however, these authors give the N^{-2} term in μ_1' incorrectly, though Haldane (1953) agrees with our result (see also Kendall and Stuart (ref. 34, Chap. 1), pp. 49–54).

2.9.2 Normal distribution with mean θ and unit variance

The density is
$$p(x, \theta) = (2\pi)^{-1/2} \exp\{-\tfrac{1}{2}(x - \theta)^2\}$$
and
$$\frac{1}{p} \frac{\partial^s p}{\partial \theta^s} = H_s(x'),$$

where $x' = x - \theta$, and $H_s(x) = \left(\exp -\tfrac{1}{2}\left(\dfrac{d}{dx}\right)^2\right) x^s$ is a Hermite polynomial. From orthogonality, $(ab) = 0$, $a \neq b$, and since alternate polynomials are of opposite parity, we have

$$(a^r b^s) = 0 \quad (ra + sb \text{ an odd number}).$$

Now $\hat{\theta} = m'$, the sample mean, so that from normality

$$\mathrm{E}\,\hat{\theta} = \theta, \qquad \mathrm{Var}\,\hat{\theta} = 1/N,$$
$$\mu_3(\hat{\theta}) = 0, \qquad \mu_4(\hat{\theta}) = 3/N^2.$$

Our expansions (2.33) should agree. In (2.33a), (2.33c), the expectations are all of odd parity, so we have $\mathrm{E}\,\hat{\theta} \sim \theta$, $\mu_3(\hat{\theta}) \sim 0$. In (2.33b),

$$I = 1, \quad b_3 = 1, \quad b_4 = 0,$$
$$c_3 = -3, \quad c_4 = -1, \quad c_5 = 3, \quad c_6 = 0, \quad c_7 = 0,$$

leading to zero N^{-2} and N^{-3} coefficients. In (2.33d),

$$e_4 = 9, \quad e_5 = 0,$$
$$f_4 = -54, \quad f_5 = -12, \quad f_6 = 45, \quad f_7 = f_8 = 0,$$

leading to $\mu_4(\hat{\theta}) \sim 3/N^2$.

It is easily seen that $\gamma_1(\hat{\theta})$ in (2.34a) is zero, and for (2.34b), $f_2 = 3$, $g_2 = -18$, $g_3 = -6$, $g_4 = 18$, $g_5 = g_6 = 0$, yielding $\gamma_2(\hat{\theta}) \sim 0$.

The logarithmic notation of (2.30) is simpler still. For

$$\frac{\partial \log p}{\partial \theta} = x', \qquad \frac{\partial^2 \log p}{\partial \theta^2} = -1,$$

$$\frac{\partial^s \log p}{\partial \theta^s} = 0, \quad s > 2.$$

Hence all entries $[a^r b^s \ldots]$ are zero if a or b or $\ldots$ exceed 2. Again,

$$[1^r 2^s] = (-1)^s [1^r]$$
$$= 0, \quad r \text{ odd}$$
$$= (-1)^s 1.3.5\ldots(r-1), \quad r \text{ even.}$$

Evidently the coefficients vanish in μ_1' and μ_3. For the variance, (2.30b) gives

$$B_1^{(2)} = 1, \quad B_2^{(2)} = -1/I + 1/I^3 = 0;$$
$$c_3^{(2)} = -3, \quad c_4^{(2)} = -1, \quad c_5^{(2)} = 3, \quad c_6^{(2)} = c_7^{(2)} = 0,$$

so $\mu_2(\hat{\theta}) \sim 1/N$.

For $\mu_4(\hat{\theta})$,

$$b_4^{(4)} = 9, \quad b_5^{(4)} = 0;$$
$$c_4^{(4)} = -54, \quad c_5^{(4)} = -12, \quad c_6^{(4)} = 45, \quad c_7^{(4)} = c_8^{(4)} = 0,$$

so $\mu_4(\hat{\theta}) \sim 3/N^2$.

For $\gamma_1(\hat{\theta})$ the coefficients are zero, and for $\gamma_2(\hat{\theta})$,

$$f_0 = -3, \quad f_2 = 3;$$
$$g_0 = 6, \quad g_2 = -18, \quad g_3 = -6, \quad g_4 = 18, \quad g_5 = g_6 = 0,$$

giving $\gamma_2(\hat{\theta}) \sim 0$.

2.9.3 Normal distribution with zero mean and variance θ

In this case,

$$p(x, \theta) = (2\pi\theta)^{-1/2} \exp(-\tfrac{1}{2}x^2/\theta) \quad (\theta > 0),$$

and the mle is $\hat{\theta} = \Sigma x_j^2/N$. Here,

$$\frac{\partial \log p}{\partial \theta} = -\tfrac{1}{2}\phi + \tfrac{1}{2}\phi^2 x^2 \quad (\phi = 1/\theta),$$

so that higher logarithmic derivatives are simple in form. We again use (2.30) and list a selection of expected values:

$[12] = -\phi^3$ $[1^23] = 7\phi^5$ $[12^2] = 5\phi^5$ $[1^22] = -9\phi^4/4$

$[1^3] = \phi^3$ $[14] = -12\phi^5$ $[23] = -7\phi^5$ $[13] = 3\phi^4$

$[2^2] = 9\phi^4/4$ $[1^33] = 49\phi^6/2$ $[1^22^2] = 137\phi^6/8$

$[2^3] = -89\phi^6/8$ $[123] = -31\phi^6/2$ $[1^24] = -57\phi^6/2$

$[24] = 57\phi^6/2$ $[3^2] = 22\phi^6$ $[15] = 60\phi^6$ $[1^4] = 15\phi^4/4$

$[1^32] = -8\phi^5$ $[12^2] = 5\phi^5$ $[1^42] = -287\phi^6/8$

$[1^33] = 49\phi^6/2$ $[1^5] = 17\phi^6$ $[2] = -\phi^2/2$ $[3] = 2\phi^3$

$[4] = -9\phi^4$ $[5] = 48\phi^5$.

Inserting the expectations in (2.30) gives

$A_1^{(1)} = A_2^{(1)} = A_3^{(1)} = 0;$

$B_1^{(2)} = 2\theta^2, \quad B_2^{(2)} = B_3^{(2)} = 0;$

$B_2^{(3)} = 8\theta^3, \quad B_3^{(3)} = 0;$

$B_2^{(4)} = 12\theta^4, \quad B_3^{(4)} = 48\theta^4, \quad B_4^{(4)} = 0.$

But $N\hat{\theta}$ is distributed as χ^2 with N degrees of freedom and characteristic function $(1 - 2i\alpha\theta)^{-N/2}$, so that $\mu_1'(\hat{\theta}) = \theta$, $\mu_2(\hat{\theta}) = 2\theta^2/N$, $\mu_3(\hat{\theta}) = 8\theta^3/N^2$, $\mu_4(\hat{\theta}) = 12\theta^4/N^2 + 48\theta^4/N^3$ in agreement.

2.9.4 Poisson distribution with mean θ

In this case,

$$p(x, \theta) = e^{-\theta}\theta^x/x! \quad (x = 0, 1, \ldots; \theta > 0)$$

and

$$\Gamma_1 = -1 + x\phi \quad (\phi = 1/\theta)$$

$$\Gamma_s = (-1)^{s-1}(s-1)! \, x\phi^s \quad (s = 2, 3, \ldots).$$

For the expectations we have

$[2] = -\phi$ $[1^4] = 3\phi^2 + \phi^3$ $[1^42] = -3\phi^3 - 11\phi^4 - \phi^5$

$[3] = 2\phi^2$ $[1^22] = -\phi^2 - \phi^3$ $[1^22^2] = \phi^3 + 5\phi^4 + \phi^5$

$[1^3] = \phi^2$ $[2^2] = \phi^2 + \phi^3$ $[2^3] = -\phi^3 - 3\phi^4 - \phi^5$

$[12] = -\phi^2$ $[1^5] = 10\phi^3 + \phi^4$ $[1^223] = -2\phi^4 - 10\phi^5 - 2\phi^6$

$[4] = -6\phi^3$ $[1^32] = -4\phi^3 - \phi^4$ $[12^3] = -3\phi^4 - 6\phi^5 - \phi^6$

$[13] = 2\phi^3$ $[1^23] = 2\phi^3 + 2\phi^4$ $[2^23] = 2\phi^4 + 6\phi^5 + 2\phi^6$

$[5] = 24\phi^4$ $[23] = -2\phi^3 - 2\phi^4$

$$[14] = -6\phi^4 \qquad [12^2] = 2\phi^3 + \phi^4$$
$$[15] = 24\phi^5 \qquad [1^3 3] = 8\phi^4 + 2\phi^5$$
$$[6] = -120\phi^5 \qquad [1^2 4] = -6\phi^4 - 6\phi^5$$
$$[16] = -120\phi^6 \qquad [123] = -4\phi^4 - 2\phi^5$$
$$[7] = 720\phi^6 \qquad [24] = 6\phi^4 + 6\phi^5$$
$$[3^2] = 4\phi^4 + 4\phi^5$$
$$[1^3 4] = -24\phi^5 - 6\phi^6$$
$$[1^2 5] = 24\phi^5 + 24\phi^6$$
$$[124] = 12\phi^5 + 6\phi^6$$
$$[13^2] = 8\phi^5 + 4\phi^6$$
$$[25] = -24\phi^5 - 24\phi^6$$
$$[34] = -12\phi^5 - 12\phi^6$$

Entering these in (2.30) gives

$$A_1^{(1)} = A_2^{(1)} = A_3^{(1)} = 0;$$
$$B_1^{(2)} = \theta, \quad B_2^{(2)} = B_3^{(2)} = 0;$$
$$B_2^{(3)} = \theta, \quad B_3^{(3)} = 0;$$
$$B_2^{(4)} = 3\theta^2, \quad B_3^{(4)} = \theta, \quad B_4^{(4)} = 0,$$

in agreement with the moments of $\hat{\theta} = m_1'$ which are $\mu_1'(\hat{\theta}) = \theta$, $\mu_2(\hat{\theta}) = \theta/N$, $\mu_3(\hat{\theta}) = \theta/N^2$, $\mu_4(\hat{\theta}) = 3\theta^2/N^2 + \theta/N^3$.

2.9.5 Fisher's "Nile" problem

In this problem, Fisher (1959, pp. 165–8) discusses what amounts to the mle of θ in the bivariate density

$$p(x, y; \theta) = \exp(-\theta x - y/\theta) \quad (0 \leq x, y < \infty, \theta > 0),$$

so that $\hat{\theta} = \sqrt{(\Sigma y / \Sigma x)}$. The density of $\hat{\theta}$, Fisher shows to be

$$f(\hat{\theta}, \theta) = \frac{2(2N-1)!}{(N-1)!\,(N-1)!} \left(\frac{\hat{\theta}}{\theta} + \frac{\theta}{\hat{\theta}}\right)^{-2N} \frac{1}{\hat{\theta}} \quad (0 < \hat{\theta} < \infty),$$

so that

$$\mathrm{E}\,\hat{\theta} = \theta(N-\tfrac{1}{2})!\,(N-\tfrac{3}{2})! \,/((N-1)!)^2$$
$$\sim \theta\left\{1 + \frac{1}{4N} + \frac{5}{32N^2} + \frac{11}{128N^3} + \frac{83}{2048N^4}\right\},$$

$$\mathrm{E}\,\hat{\theta}^2 = \theta^2 N/(N-1),$$
$$\mathrm{E}\,\hat{\theta}^3 = \theta^3 (N-\tfrac{5}{2})!\,(N+\tfrac{1}{2})!\,/\,((N-1)!)^2$$
$$\sim \theta^3 \left\{ 1 + \frac{9}{4N} + \frac{117}{32N^2} + \frac{723}{128N^3} + \frac{17523}{2048N^4} \right\},$$
$$\mathrm{E}\,\hat{\theta}^4 = \theta^4 (N+1)N\,/\,((N-1)(N-2)).$$

Applying correction formulae, we have

$$\mathrm{Var}\,(\hat{\theta}/\theta) \sim \frac{1}{2N} + \frac{5}{8N^2} + \frac{3}{4N^3},$$

$$\mu_3(\hat{\theta}/\theta) \sim \frac{3}{4N^2} + \frac{35}{16N^3},$$

$$\mu_4(\hat{\theta}/\theta) \sim \frac{3}{4N^2} + \frac{33}{8N^3} + \frac{927}{64N^4}.$$

Using our asymptotic expansions (2.30),

$$\Gamma_1 = -x + y\phi, \qquad (\phi = 1/\theta)$$
$$\Gamma_s = (-1)^{s+1} s!\, y\phi^{s+1}, \quad s = 2, 3, \ldots.$$

Moreover,

$[2] = -2\phi^2$	$[3] = 6\phi^3$	$[4] = -24\phi^4$
$[5] = 120\phi^5$	$[6] = -720\phi^6$	$[7] = 5040\phi^7$
$[12] = -2\phi^3$	$[1^2 3] = 24\phi^5$	$[12^2] = 16\phi^5$
$[1^2 2] = -8\phi^4$	$[14] = -24\phi^5$	$[23] = -24\phi^5$
$[13] = 6\phi^4$	$[2^2] = 8\phi^4$	$[1^3 3] = 72\phi^6$
$[1^2 2^2] = 64\phi^6$	$[1^2 4] = -96\phi^6$	$[123] = -48\phi^6$
$[2^3] = -48\phi^6$	$[24] = 96\phi^6$	$[3^2] = 72\phi^7$
$[1^3] = 0$	$[1^5] = 0$	$[15] = 120\phi^6$
$[1^3 2] = -24\phi^4$	$[1^4 2] = -144\phi^6$	$[1^3 4] = -288\phi^7$
$[1^2 23] = -192\phi^7$	$[12^3] = -144\phi^7$	$[1^2 5] = 480\phi^7$
$[124] = 192\phi^7$	$[2^2 3] = 144\phi^7$	$[13^2] = 144\phi^7$
$[16] = -720\phi^7$	$[25] = -480\phi^7$	$[34] = -288\phi^7$
$[1^4] = 24\phi^4.$		

From these,

$$A_1^{(1)} = \theta/4, \quad A_2^{(1)} = 5\theta/32, \quad A_3^{(1)} = 11\theta/128;$$
$$B_1^{(2)} = \theta^2/2, \quad B_2^{(2)} = 5\theta^2/8, \quad B_3^{(2)} = 3\theta^2/4;$$
$$B_2^{(3)} = 3\theta^3/4, \quad B_3^{(3)} = 35\theta^3/16;$$
$$B_2^{(4)} = 3\theta^4/4, \quad B_3^{(4)} = 33\theta^4/8, \quad B_4^{(4)} = 927\theta^4/64,$$

in exact agreement with the results derived from the true distribution of $\hat{\theta}$.

REFERENCES

1. Bartlett, M.S. (1952). "Approximate confidence intervals, II." *Biometrika*, **40**, 306–17.
2. Bowman, K.O. (1963). "Moments to higher orders for maximum likelihood estimators, with an application to the negative binomial distribution". Thesis in support of the Ph.D. degree at Virginia Polytechnic Institute.
3. Cox, D.R. and Snell, E.J. (1968). "A general definition of residuals". *J. Roy. Statist. Soc. (B)*, **30**, 248–75.
4. Fisher, Sir Ronald A. (1959). *Statistical Methods and Scientific Inference*, Hafner, New York.
5. Haldane, J.B.S. (1953). "The estimation of two parameters from a sample". *Sankhyā*, **12**, 313–20.
6. Haldane, J.B.S. and Smith, S.M. (1956). "The sampling distribution of a maximum likelihood estimate". *Biometrika*, **43**, 96–103.
7. Shenton, L.R. (1959). "The distribution of moment estimators". *Biometrika*, **46**, 296–305.
8. Shenton, L.R. and Bowman, K.O. (1963). "Higher moments of a maximum likelihood estimator". *J. Roy. Statist. Soc. (B)*, **25**, 305–17.
9. Shenton, L.R. and Wallington, P.A. (1962). "The bias of moment estimators with an application to the negative binomial distribution". *Biometrika*, **49**, 193.

3 BIAS AND COVARIANCE IN MULTIPARAMETER ESTIMATION[*]

3.1 Introduction

If there are h parameters involved in the probabilities $p_j(\theta)$ for the cell frequencies, there are now h likelihood equations which can be written in summary notation:

$$L_\alpha + x_r L_{r\alpha} + \tfrac{1}{2} x_s x_t L_{st\alpha} + \ldots = 0 \quad (\alpha, r, s, t = 1, 2, \ldots, h), \quad (3.1)$$

where $L_\alpha = n_r \hat{\Gamma}^r_\alpha$, $\hat{\Gamma}^r_\alpha = d\log p_r(\hat{\theta})/d\hat{\theta}_\alpha$, $x_r = \hat{\theta}_r - \theta_r$. The inversion of these equations using multivariate Lagrange (see, for example, Good, 1959) presents formidable complexities. The indirect approach using (n_r) as argument in the Taylor expansion along with logarithmic derivatives will be described first; the alternate form in terms of derivatives of (p_r) has not been developed because of its limited utility. Mention will be made of the approach to the moments of $\hat{\theta}$ using the adjusted order of magnitude method of section 2.6.

3.2 Notation

Summatory expressions over classes involving log-likelihood derivatives are dealt with as follows:

$$(p_r \Gamma^r_{\alpha_1 \alpha_2 \ldots \alpha_l} \Gamma^r_{\beta_1 \beta_2 \ldots \beta_m} \Gamma^r_{\gamma_1 \gamma_2 \ldots \gamma_n})$$
$$= (p \Gamma_{\alpha_1 \alpha_2 \ldots \alpha_l} \Gamma_{\beta_1 \beta_2 \ldots \beta_m} \Gamma_{\gamma_1 \gamma_2 \ldots \gamma_n})$$
$$= [\alpha_1 \alpha_2 \ldots \alpha_l, \beta_1 \beta_2 \ldots \beta_m, \gamma_1 \gamma_2 \ldots \gamma_n].$$

For example,

$$[\alpha, \beta, \gamma] = (p\Gamma_\alpha \Gamma_\beta \Gamma_\gamma) = \sum_r p_r \frac{\partial \log p_r}{\partial \theta_\alpha} \frac{\partial \log p_r}{\partial \theta_\beta} \frac{\partial \log p_r}{\partial \theta_\gamma}$$

$$[\alpha\beta, \gamma] = (p\Gamma_{\alpha\beta} \Gamma_\gamma) = \sum_r p_r \frac{\partial^2 \log p_r}{\partial \theta_\alpha \partial \theta_\beta} \frac{\partial \log p_r}{\partial \theta_\gamma}$$

---

[*] The reader who wishes to use the formulae without the distraction of tiresome details should look at section 3.2 for the notation, and equations (3.8), (3.9b) and (3.11) for the covariance, and (3.13) for the bias. The remainder of the chapter is concerned with the derivation of the formulae and validation, using special cases.

$$[\alpha\beta\gamma] = (p\Gamma_{\alpha\beta\gamma}) = \sum_r p_r \frac{\partial^3 \log p_r}{\partial\theta_\alpha \partial\theta_\beta \partial\theta_\gamma}.$$

Occasionally, in the special cases considered, we meet expressions such as

$$\sum_r p_r \left(\frac{\partial^2 \log p_r}{\partial\theta_1^2}\right)^2$$

which should be written as $(p_r \Gamma_{11}^r \Gamma_{11}^r)$; we abbreviate this to $(p\Gamma_{11}^2)$. It is generally clear from the context whether the superscript in Γ represents a power or a summatory symbol.

3.3 Taylor expansion

Since there are h parameters, we now have

$$\hat{\theta}_a = \theta_a + \phi_1^a + \frac{1}{2!}\phi_2^a + \frac{1}{3!}\phi_3^a + \ldots \quad (a = 1, 2, \ldots h), \tag{3.2}$$

where

(i) $\phi_1^a = \epsilon_r \dfrac{\bar{\partial}\theta_a}{\partial n_r}, \quad \phi_2^a = \epsilon_r \epsilon_s \dfrac{\bar{\partial}^2 \theta_a}{\partial n_r \partial n_s},$ etc.;

(ii) $\epsilon_r = n_r - p_r, \quad E\epsilon_r = 0;$

(iii) $\dfrac{\bar{\partial}\theta_a}{\partial n_r} = \dfrac{\partial\hat{\theta}_a}{\partial n_r}\bigg|_{n_r = p_r, \, \hat{\theta} = \theta}$

and so on.

The derivatives in (3.2) are found indirectly from

$$n_r \hat{\Gamma}_\alpha^r = 0 \quad (\alpha = 1, 2, \ldots, h) \tag{3.3}$$

by successive differentiation with respect to $n_r, n_s, \ldots$. Thus

$$\hat{\Gamma}_\alpha^r + (n_r \hat{\Gamma}_{\alpha\beta}^r)\frac{\partial\hat{\theta}_\beta}{\partial n_r} = 0, \tag{3.4a}$$

from which

$$\frac{\bar{\partial}\theta_\alpha}{\partial n_r} = L^{\alpha\beta}\Gamma_\beta^r, \tag{3.4b}$$

where

(i) $L_{\alpha\beta} = -(p\Gamma_{\alpha\beta}) = (p\Gamma_\alpha \Gamma_\beta)$

(ii) $L_{\alpha\beta} L^{\beta\gamma} = \epsilon_{\alpha\gamma}$ (Kronecker delta),

so that $L^{\beta\alpha}$ refers to an element in the inverse matrix $[\mathbf{L}_{\alpha\beta}]^{-1}$, where $[\mathbf{L}_{\alpha\beta}]$ has h rows and columns and is assumed nonsingular.

Differentiating (3.4a) with respect to n_s,

$$\hat{\Gamma}_{\alpha\beta}^r \frac{\partial \hat{\theta}_\beta}{\partial n_s} + \hat{\Gamma}_{\alpha\beta}^s \frac{\partial \hat{\theta}_\beta}{\partial n_r} + (n_r \hat{\Gamma}_{\alpha\beta\gamma}) \frac{\partial \hat{\theta}_\beta}{\partial n_r} \frac{\partial \hat{\theta}_\gamma}{\partial n_s} + (n_r \hat{\Gamma}_{\alpha\beta}^r) \frac{\partial^2 \hat{\theta}_\beta}{\partial n_r \partial n_s} = 0, \quad (3.5a)$$

from which

$$L_{\alpha\beta} \frac{\bar{\partial}^2 \theta_\beta}{\partial n_r \partial n_s} = L^{\beta\gamma} \Gamma_{\alpha\beta}^r \Gamma_\gamma^s + L^{\beta\gamma} \Gamma_{\alpha\beta}^s \Gamma_\gamma^r + [\alpha\beta\gamma] L^{\beta\delta} L^{\gamma\epsilon} \Gamma_\delta^r \Gamma_\epsilon^s. \quad (3.5b)$$

It will be seen that since partial differentiation is usually commutative, the right-hand member of (3.5b) should also have this property with respect to r, s. This is readily verified; in fact, the last member of (3.5b) under $r \leftrightarrow s$ is invariant, using $\delta \leftrightarrow \epsilon, \gamma \leftrightarrow \beta$. For the third derivative we find

$$-(n\hat{\Gamma}_{\alpha\beta}) \frac{\partial^3 \hat{\theta}_\beta}{\partial n_r \partial n_s \partial n_t} = a_1 + a_2 + a_3 + a_4, \quad (3.6a)$$

where

$$a_1 = \hat{\Gamma}_{\alpha\beta\gamma}^r \frac{\partial \hat{\theta}_\beta}{\partial n_s} \frac{\partial \hat{\theta}_\gamma}{\partial n_t} + \hat{\Gamma}_{\alpha\beta\gamma}^s \frac{\partial \hat{\theta}_\beta}{\partial n_r} \frac{\partial \hat{\theta}_\gamma}{\partial n_t} + \hat{\Gamma}_{\alpha\beta\gamma}^t \frac{\partial \hat{\theta}_\beta}{\partial n_r} \frac{\partial \hat{\theta}_\gamma}{\partial n_s},$$

$$a_2 = \hat{\Gamma}_{\alpha\beta}^r \frac{\partial^2 \hat{\theta}_\beta}{\partial n_s \partial n_t} + \hat{\Gamma}_{\alpha\beta}^s \frac{\partial^2 \hat{\theta}_\beta}{\partial n_r \partial n_t} + \hat{\Gamma}_{\alpha\beta}^t \frac{\partial^2 \hat{\theta}_\beta}{\partial n_r \partial n_s},$$

$$a_3 = (n\hat{\Gamma}_{\alpha\beta\gamma}) \left[\frac{\partial^2 \hat{\theta}_\gamma}{\partial n_s \partial n_t} \frac{\partial \hat{\theta}_\beta}{\partial n_r} + \frac{\partial^2 \hat{\theta}_\beta}{\partial n_r \partial n_t} \frac{\partial \hat{\theta}_\gamma}{\partial n_s} + \frac{\partial^2 \hat{\theta}_\beta}{\partial n_r \partial n_s} \frac{\partial \hat{\theta}_\gamma}{\partial n_t} \right],$$

$$a_4 = (n\hat{\Gamma}_{\alpha\beta\gamma\delta}) \frac{\partial \hat{\theta}_\beta}{\partial n_r} \frac{\partial \hat{\theta}_\gamma}{\partial n_s} \frac{\partial \hat{\theta}_\delta}{\partial n_t},$$

from which

$$L_{\alpha\beta} \frac{\bar{\partial}^3 \theta_\beta}{\partial n_r \partial n_s \partial n_t} = \underline{3}\Gamma_{\alpha\beta\gamma}^r \frac{\bar{\partial}\theta_\beta}{\partial n_s} \frac{\bar{\partial}\theta_\gamma}{\partial n_t} + \underline{3}\Gamma_{\alpha\beta}^r \frac{\bar{\partial}^2 \theta_\beta}{\partial n_s \partial n_t}$$

$$+ \underline{3}[\alpha\beta\gamma] \frac{\bar{\partial}\theta_\beta}{\partial n_r} \frac{\bar{\partial}^2 \theta_\gamma}{\partial n_s \partial n_t} + [\alpha\beta\gamma\delta] \frac{\bar{\partial}\theta_\beta}{\partial n_r} \frac{\bar{\partial}\theta_\gamma}{\partial n_s} \frac{\bar{\partial}\theta_\delta}{\partial n_t}, \quad (3.6b)$$

where the underlined coefficients are used to indicate the total number of terms of the permutation of r, s, t (or its equivalent) involved in each term. Thus the second member of (3.6b) refers to the *three* terms in a_1 of (3.6a); when ultimately summed in expectation, each of these terms has the same value. A similar remark applies to the terms in a_2, etc. With this convention, differentiating (3.6a) with respect to n_u and reverting to population values yields

$$L_{\alpha\beta} \frac{\bar{\partial}^4 \theta_\beta}{\partial n_r \partial n_s \partial n_t \partial n_u} = \underline{4}\Gamma_{\alpha\beta\gamma\delta}^r \frac{\bar{\partial}\theta_\beta}{\partial n_s} \frac{\bar{\partial}\theta_\gamma}{\partial n_t} \frac{\bar{\partial}\theta_\delta}{\partial n_u}$$

$$+ \underline{6}[\alpha\beta\gamma\delta] \frac{\bar{\partial}^2 \theta_\gamma}{\partial n_s \partial n_u} \frac{\bar{\partial}\theta_\beta}{\partial n_r} \frac{\bar{\partial}\theta_\delta}{\partial n_t} + \underline{3}[\alpha\beta\gamma] \frac{\bar{\partial}^2 \theta_\beta}{\partial n_r \partial n_u} \frac{\bar{\partial}^2 \theta_\gamma}{\partial n_s \partial n_t} +$$

$$+ 12\Gamma^r_{\alpha\beta\gamma} \frac{\bar{\partial}\theta_\gamma}{\partial n_t} \frac{\bar{\partial}^2\theta_\beta}{\partial n_s \partial n_u} + 4[\alpha\beta\gamma] \frac{\bar{\partial}^3\theta_\gamma}{\partial n_s \partial n_t \partial n_u} \frac{\bar{\partial}\theta_\beta}{\partial n_r}$$

$$+ 4\Gamma^r_{\alpha\beta} \frac{\bar{\partial}^3\theta_\beta}{\partial n_s \partial n_t \partial n_u} + [\alpha\beta\gamma\delta\nu] \frac{\bar{\partial}\theta_\gamma}{\partial n_s} \frac{\bar{\partial}\theta_\beta}{\partial n_r} \frac{\bar{\partial}\theta_\delta}{\partial n_t} \frac{\bar{\partial}\theta_\nu}{\partial n_u}. \tag{3.7}$$

3.4 Covariances

From (3.2),

$$E(\hat{\theta}_a - \theta_a)(\hat{\theta}_b - \theta_b) = \psi_1(a,b) + \psi_2(a,b) + \ldots \tag{3.8}$$

where ψ_1 refers to the N^{-1} coefficient, ψ_2 to N^{-2} and so on. Thus

$$\psi_1 = E \phi_1^a \phi_1^b \tag{3.9a}$$
$$= E \epsilon_r \epsilon_s \Gamma^r_\beta \Gamma^s_\gamma L^{a\beta} L^{b\gamma}$$
$$= (p\Gamma_\beta \Gamma_\gamma) L^{a\beta} L^{b\gamma} \quad ((p\Gamma_\beta) = (p\Gamma_\gamma) = 0),$$

from equation (2.5b). But $L_{\beta\gamma} L^{a\beta} = \epsilon_{a\gamma}$, so

$$\psi_1 = L^{ab}, \tag{3.9b}$$

as expected.

For the N^{-2} covariances,

$$\psi_2(a,b) = (6A_{12} + 2A_{13} + 3A_{22})/12, \tag{3.9c}$$

where
$$A_{12} = E_2(\phi_1^a \phi_2^b + \phi_2^a \phi_1^b),$$
$$A_{13} = E_2(\phi_1^a \phi_3^b + \phi_3^a \phi_1^b),$$
$$A_{22} = E_2 \phi_2^a \phi_2^b.$$

From (3.4b) and (3.5b),

$$E_2 \phi_1^a \phi_2^b = E_2 \epsilon_r \epsilon_s \epsilon_t L^{a\beta} L^{\alpha b} \{L^{\gamma \zeta} \Gamma^r_\beta \Gamma^s_\gamma \Gamma^t_{\alpha\zeta} + L^{\gamma\zeta} \Gamma^r_\beta \Gamma^s_{\alpha\zeta} \Gamma^t_\gamma$$
$$+ [\alpha\gamma\zeta] L^{\delta\zeta} L^{\gamma\epsilon} \Gamma^r_\beta \Gamma^s_\epsilon \Gamma^t_\delta\}$$
$$= 2L^{a\beta} L^{\alpha b} L^{\gamma\zeta} [\beta, \gamma, \alpha\zeta] + 2L^{ab} + L^{a\beta} L^{\alpha b} L^{\delta\zeta} L^{\gamma\epsilon} [\alpha\gamma\zeta] [\beta, \delta, \epsilon]. \tag{3.10a}$$

Similarly, from (3.4b) and (3.6b),

$$E_2 \phi_1^a \phi_3^b = \sum_{n=1}^{4} \psi_n, \tag{3.10b}$$

where
$$\psi_1 = 3E_2 \epsilon_r \epsilon_s \epsilon_t \epsilon_u L^{a\epsilon} L^{\alpha b} L^{\beta\zeta} L^{\gamma\eta} \Gamma^r_{\alpha\beta\gamma} \Gamma^s_\zeta \Gamma^t_\eta \Gamma^u_\epsilon$$
$$= 3L^{a\epsilon} L^{\alpha b} L^{\gamma\beta} [\alpha\beta\gamma, \epsilon] + 6L^{a\gamma} L^{\alpha b} L^{\beta\zeta} [\alpha\beta\gamma, \zeta],$$

$$\psi_2 = 3E_2 \epsilon_r \epsilon_s \epsilon_t \epsilon_u L^{a\epsilon} L^{\alpha b} \Gamma^u_\epsilon \Gamma^r_{\alpha\beta} \{2L^{\zeta\beta} L^{\eta\theta} \Gamma^s_{\zeta\eta} \Gamma^t_\theta + [\zeta\eta\theta] L^{\zeta\beta} L^{\theta\kappa} L^{\eta\lambda} \Gamma^s_\kappa \Gamma^t_\lambda\},$$

which, after simplifying the expectations, gives

$$\psi_2 = -6L^{ab} + 6L^{a\epsilon}L^{\alpha b}L^{\zeta\beta}[\alpha\beta, \zeta\epsilon] + L^{a\epsilon}L^{\alpha b}L^{\zeta\beta}L^{\eta\theta}$$
$$\times \{6[\alpha\beta, \epsilon][\zeta\eta, \theta] + 6[\alpha\beta, \theta][\zeta\eta, \epsilon] + 3[\zeta\eta\theta][\alpha\beta, \epsilon] + 6[\zeta\eta\epsilon][\alpha\beta, \theta]\}.$$

(3.10c)

Similarly for the remaining terms, which correspond to the last two terms in (3.6b), we have,

$$\psi_3 = 3[\alpha\beta\gamma\delta]L^{\alpha b}L^{a\beta}L^{\gamma\delta},$$

$$\psi_4 = L^{\alpha b}L^{a\gamma}L^{\beta\eta}L^{\theta i}\{6[\alpha\beta\gamma][\eta\theta, i] + 6[\alpha\beta i][\eta\theta, \gamma]$$
$$+ 6[\alpha\beta i][\eta\gamma, \theta] + 3[\alpha\beta\gamma][\eta\theta i] + 6[\alpha\beta\theta][\eta\gamma i]\}.$$

From (3.5b) we have

$$A_{22} = -4L^{ab} + 4L^{a\alpha}L^{b\eta}L^{\beta\theta}[\alpha\beta, \eta\theta]$$
$$+ L^{\alpha\beta}L^{\delta\zeta}L^{\gamma\epsilon}L^{b\alpha}\{[\beta\epsilon\gamma][\alpha\delta\zeta] + 2[\beta\delta\gamma][\alpha\epsilon\zeta] + 2[\beta\epsilon\gamma][\alpha\delta, \zeta]$$
$$+ 4[\beta\delta\gamma][\alpha\zeta, \epsilon] + 2[\alpha\delta\zeta][\beta\epsilon, \gamma] + 4[\alpha\delta\gamma][\beta\epsilon, \zeta]$$
$$+ 4[\alpha\delta, \zeta][\beta\epsilon, \gamma] + 4[\beta\epsilon, \zeta][\alpha\delta, \gamma]\}.$$

(3.10d)

Hence, returning to (3.9c) with contributions from (3.10a), (3.10b) and (3.10d), and using the interchange $a \leftrightarrow b$, we have for the N^{-2} term in a covariance of any two estimators, in terms involving equal powers of elements in $[L_{\alpha\beta}]^{-1}$,

$$\text{Cov}_2(\hat{\theta}_a, \hat{\theta}_b) = \theta_1 + \theta_3 + \theta_4,$$

(3.11)

where

$$\theta_1 = -L^{ab},$$

$$\theta_3 = L^{a\beta}L^{b\alpha}L^{\gamma\delta}\{[\alpha\delta, \beta, \gamma] + [\beta\delta, \alpha, \gamma] + [\alpha\beta\gamma\delta]$$
$$+ 3[\alpha\delta, \gamma\beta] + 2[\alpha\beta\gamma, \delta] + \tfrac{1}{2}[\beta\gamma\delta, \alpha] + \tfrac{1}{2}[\alpha\gamma\delta, \beta]\},$$

$$\theta_4 = L^{a\beta}L^{b\alpha}L^{\gamma\epsilon}L^{\delta\zeta}\{\tfrac{1}{2}[\alpha\gamma\zeta][\beta, \delta, \epsilon] + \tfrac{1}{2}[\beta\gamma\zeta][\alpha, \delta, \epsilon]$$
$$+ [\alpha\gamma\beta][\delta\epsilon\zeta] + 5[\alpha\gamma\delta][\beta\epsilon\zeta]/2 + [\alpha, \delta\epsilon][\beta\zeta\gamma] + [\beta, \delta\epsilon][\alpha\zeta\gamma]$$
$$+ 2[\alpha\beta\zeta][\delta\gamma, \epsilon] + 3[\alpha\zeta, \epsilon][\beta\gamma\delta] + 3[\alpha\gamma\delta][\beta\zeta, \epsilon]$$
$$+ \tfrac{1}{2}[\alpha, \beta\zeta][\gamma\delta\epsilon] + \tfrac{1}{2}[\alpha\zeta, \beta][\gamma\delta\epsilon] + [\alpha\delta, \epsilon][\gamma\zeta, \beta]$$
$$+ [\beta\delta, \epsilon][\gamma\zeta, \alpha] + [\alpha\delta, \beta][\gamma\zeta, \epsilon] + [\beta\delta, \alpha][\gamma\zeta, \epsilon]$$
$$+ [\alpha\delta, \gamma][\beta\epsilon, \zeta]\}.$$

3.5 Bias

From (3.2), for the N^{-1} bias,

$$E_1 \hat{\theta}_a = \tfrac{1}{2} E_1 \phi_2^a$$

$$= \tfrac{1}{2} E_1 \epsilon_r \epsilon_s \frac{\bar{\partial}^2 \hat{\theta}_a}{\partial n_r \partial n_s}. \qquad (3.12a)$$

But from equations (3.5b) and (2.5b),

$$\frac{\bar{\partial}^2 \theta_\beta}{\partial n_r \partial n_s} = 2L^{\beta\gamma}(p\Gamma_{\alpha\beta}\Gamma_\gamma) + [\alpha\beta\gamma]L^{\beta\delta}L^{\gamma\epsilon}(p\Gamma_\delta\Gamma_\epsilon),$$

so that, applying the symbol $L^{\alpha a}$,

$$E_1 \epsilon_r \epsilon_s \frac{\bar{\partial}^2 \hat{\theta}_a}{\partial n_r \partial n_s} = L^{a\alpha} L^{\beta\gamma}\{2[\alpha\beta, \gamma] + [\alpha\beta\gamma]\},$$

from which[*]

$$E_1 \hat{\theta}_a = L^{a\alpha} L^{\beta\gamma}\{[\alpha\beta, \gamma] + \tfrac{1}{2}[\alpha\beta\gamma]\}. \qquad (3.12b)$$

An alternative to (3.12b) is to use (3.5b) to derive

$$E_1 L_{\alpha\beta} \epsilon_r \epsilon_s \frac{\bar{\partial}^2 \hat{\theta}_\beta}{\partial n_r \partial n_s} = L^{\beta\gamma}\{2[\alpha\beta, \gamma] + [\alpha\beta\gamma]\},$$

from which, writing $E_1(\hat{\theta}_\alpha - \theta_\alpha) = b_\alpha$, we have

$$L_{\alpha\beta} b_\beta = L^{\beta\gamma}\{[\alpha\beta, \gamma] + \tfrac{1}{2}[\alpha\beta\gamma]\}. \qquad (3.12c)$$

In some cases this may be simpler to handle than (3.12b).

Example 3.1   In single-parameter estimation show that

$$E_1(\hat{\theta} - \theta) = \frac{[12] + \tfrac{1}{2}[3]}{[2]^2}$$

$$= -\frac{E\left(\dfrac{1}{p^2} \dfrac{\partial p}{\partial \theta} \dfrac{\partial^2 p}{\partial \theta^2}\right)}{2I^2},$$

where $I = -(p\Gamma_2) = (p\Gamma_1^2)$.

Example 3.2   Show that the biases b_1, b_2 in the simultaneous estimation of θ_1, θ_2 are given by

---

[*] Haldane (1953) gave the bias in the two-parameter case. See also the references to (2.10) in **2.5**.

$$b_1 P_{1,1} + b_2 P_{1,2} = -\frac{(P_{2,2}P_{1,11} - 2P_{1,2}P_{1,12} + P_{1,1}P_{1,22})}{2(P_{1,1}P_{2,2} - P_{1,2}^2)}$$

$$b_1 P_{2,1} + b_2 P_{2,2} = -\frac{(P_{2,2}P_{2,11} - 2P_{1,2}P_{2,12} + P_{1,1}P_{2,22})}{2(P_{1,1}P_{2,2} - P_{1,2}^2)}$$

(Shenton and Wallington, 1962)

Example 3.3 Show that formula (3.12c) is equivalent to

$$P_{\alpha,\beta} b_\beta = -\tfrac{1}{2} P_{\alpha,\beta\gamma} P^{\beta,\gamma},$$

where the probability of the variate x is $P(x; \theta_1, \theta_2, \ldots, \theta_h)$, and

$$P_{\alpha,\beta} = E\left(\frac{1}{P^2} \frac{\partial P}{\partial \theta_\alpha} \frac{\partial P}{\partial \theta_\beta}\right),$$

$$P_{\alpha,\beta\gamma} = E\left(\frac{1}{P^2} \frac{\partial P}{\partial \theta_\alpha} \frac{\partial^2 P}{\partial \theta_\beta \partial \theta_\gamma}\right)$$

$[P_{\alpha,\beta}][P^{\beta,\alpha}] = \mathbf{I}$ (a unit matrix of order h).

(Shenton and Wallington, 1962)

Example 3.4 For the ml estimators of $\theta_1 = \mu$, $\theta_2 = \sigma$ for the distribution $N(\mu, \sigma^2)$, show that

$$P_{1,1} = 1/\sigma^2 \qquad P_{1,2} = 0 \qquad P_{2,2} = 2/\sigma^2$$
$$P_{1,11} = 0 \qquad P_{1,12} = 0 \qquad P_{1,22} = 0$$
$$P_{2,11} = 2/\sigma^3 \qquad P_{2,12} = 0 \qquad P_{2,22} = 2/\sigma^3.$$

Deduce that $b_2 = E_1(\hat{\theta}_2 - \theta_2) = -3\sigma/4$.

Example 3.5 For the bivariate normal density $P = k \exp\{-A(x,y)\}$, where $k^{-1} = 2\pi \sigma_1 \sigma_2 \sqrt{(1-\rho^2)}$ and

$$A(x,y) = (X^2 - 2\rho XY + Y^2)/2(1-\rho^2)$$

$$(X = (x-\alpha)/\sigma_1, \quad Y = (y-\beta)/\sigma_2),$$

define $\theta_1 = \alpha$, $\theta_2 = \beta$, $\theta_3 = \sigma_1$, $\theta_4 = \sigma_2$, $\theta_5 = \rho$ $\begin{pmatrix} \sigma_1, \sigma_2 > 0 \\ |\rho| < 1 \end{pmatrix}$.

Show that

(a) $P_{1,1} = \delta/\theta_3^2 \qquad P_{1,2} = -\delta\theta_5/\theta_3\theta_4 \qquad P_{1,r} = 0 \quad (r = 3, 4, 5);$

$P_{2,2} = \delta/\theta_4^2 \qquad P_{2,r} = 0 \quad (r = 3, 4, 5);$

$P_{3,3} = (2-\theta_5^2)\delta/\theta_3^2 \qquad P_{3,4} = -\delta\theta_5^2/\theta_3\theta_4 \qquad P_{3,5} = -\delta\theta_5/\theta_3$

$P_{4,4} = (2-\theta_5^2)\delta/\theta_4^2 \qquad P_{4,5} = -\delta\theta_5/\theta_4 \qquad P_{5,5} = (1+\theta_5^2)\delta^2$

$(\delta^{-1} = 1 - \theta_5^2);$

(b) $P_{1,\alpha\beta} = P_{2,\alpha\beta} = 0 \quad (\alpha, \beta = 1, 2, \ldots, 5)$.

For the inverse matrix, show that

$$P^{1,1} = \theta_3^2 \qquad P^{1,2} = \theta_3\theta_4\theta_5 \qquad P^{1,r} = 0 \quad (r = 3, 4, 5);$$
$$P^{2,2} = \theta_4^2 \qquad P^{2,r} = 0 \quad (r = 3, 4, 5);$$
$$P^{3,3} = \tfrac{1}{2}\theta_3^2 \qquad P^{3,4} = \tfrac{1}{2}\theta_3\theta_4\theta_5^2 \qquad P^{3,5} = \tfrac{1}{2}\theta_3\theta_5/\delta;$$
$$P^{4,4} = \tfrac{1}{2}\theta_4^2 \qquad P^{4,5} = \tfrac{1}{2}\theta_4\theta_5/\delta \qquad P^{5,5} = 1/\delta^2.$$

If $R_\alpha = P_{\alpha,\beta\gamma}P^{\beta,\gamma}$, show that

$$R_1 = R_2 = 0 \quad R_3 = (3 - \theta_5^2)/\theta_3 \quad R_4 = (3 - \theta_5^2)/\theta_4 \quad R_5 = (\theta_5^3 - 2\theta_5)\delta.$$

Deduce that $b_1 = b_2 = 0$ and

$$2\begin{bmatrix} b_3 \\ b_4 \\ b_5 \end{bmatrix} = -\begin{bmatrix} \tfrac{1}{2}\theta_3^2 & \tfrac{1}{2}\theta_3\theta_4\theta_5^2 & \tfrac{1}{2}\theta_3\theta_5/\delta \\ \tfrac{1}{2}\theta_3\theta_4\theta_5^2 & \tfrac{1}{2}\theta_4^2 & \tfrac{1}{2}\theta_4\theta_5/\delta \\ \tfrac{1}{2}\theta_3\theta_5/\delta & \tfrac{1}{2}\theta_4\theta_5/\delta & 1/\delta^2 \end{bmatrix}\begin{bmatrix} (3-\theta_5^2)/\theta_3 \\ (3-\theta_5^2)/\theta_4 \\ (\theta_5^3 - 2\theta_5)\delta \end{bmatrix}.$$

(Shenton and Wallington, 1962)

For the N^{-2} bias we have to take into account ϕ_3^a and ϕ_4^a in (3.2), which in turn require the manipulation of terms in (3.6b) and (3.7); for example, (3.7) involves (3.6b), (3.5b) and (3.4b). After appropriate substitutions and simplifications (noting, for example, the equivalence of permuted terms such as those in a_1, a_2, a_3 in (3.6a)), we find

$$\mathrm{E}_2\hat{\theta}_a = L^{a\alpha}L^{\beta\gamma}B_2 + L^{a\alpha}L^{\beta\delta}L^{\gamma\epsilon}B_3 + L^{a\alpha}L^{\beta\gamma}L^{\delta\zeta}L^{\epsilon\eta}B_4$$
$$+ L^{a\alpha}L^{\beta\gamma}L^{\delta\epsilon}L^{\zeta\eta}L^{\theta i}B_5, \tag{3.13}$$

where

$$2B_2 = -[\alpha\beta\gamma] - 2[\alpha\beta, \gamma],$$
$$8B_3 = [\alpha\beta\gamma\delta\epsilon] + 4[\alpha\gamma, \beta\delta\epsilon] + 8[\delta\epsilon, \alpha\beta\gamma] + 4[\alpha\beta\gamma\epsilon, \delta] + 4[\alpha\beta\gamma, \delta, \epsilon]$$
$$+ 8[\alpha\beta, \gamma\delta, \epsilon],$$
$$4B_4 = (2[\alpha\zeta\epsilon\delta][\eta\beta, \gamma] + 2[\beta\epsilon\delta\zeta][\alpha\gamma, \eta] + 4[\alpha\beta\epsilon\delta][\eta\zeta, \gamma])$$
$$+ ([\alpha\zeta\epsilon\delta][\eta\gamma\beta] + 2[\alpha\beta\epsilon\delta][\eta\gamma\zeta] + 2[\delta\beta\epsilon\eta][\alpha\gamma\zeta])$$
$$+ (2[\alpha\epsilon\delta][\eta\beta, \zeta\gamma] + 4[\alpha\gamma\zeta][\delta\eta, \epsilon\beta] + 4[\beta\epsilon\delta][\alpha\gamma, \eta\zeta]$$
$$+ 2[\zeta\gamma\beta][\alpha\eta, \epsilon\delta]) + (4[\alpha\zeta\epsilon, \eta][\delta\beta, \gamma] + 4[\alpha\zeta\epsilon, \gamma][\delta\beta, \eta]$$
$$+ 4[\delta\beta\epsilon, \eta][\alpha\zeta, \gamma]) + (2[\alpha\beta\epsilon, \eta][\gamma\delta\zeta] + 4[\delta\beta\epsilon, \eta][\alpha\gamma\zeta]$$
$$+ 4[\alpha\beta\epsilon, \zeta][\gamma\delta\eta] + 2[\delta\eta\epsilon, \beta][\alpha\gamma\zeta]) + (4[\alpha\eta, \zeta\gamma][\epsilon\delta, \beta]$$
$$+ 4[\epsilon\delta, \zeta\gamma][\alpha\eta, \beta] + 4[\alpha\eta, \epsilon\delta][\zeta\gamma, \beta]) + (4[\alpha\gamma\eta][\epsilon\zeta, \beta, \delta] +$$

$$+ 2[\gamma\delta\epsilon][\alpha\beta,\zeta,\eta]) + 2[\alpha\beta\delta\epsilon][\gamma,\zeta,\eta]/3,$$

$$8B_5 = [\alpha\delta\zeta]\{[\epsilon\beta\gamma][\eta\theta i] + 2[\epsilon\theta\gamma][\eta\beta i] + 4[\eta\beta\epsilon][\theta\gamma i] + 8[\eta\beta\theta][\epsilon\gamma i]$$
$$+ 2[\epsilon\beta\gamma][\eta\theta,i] + 4[\epsilon\theta\gamma][\eta\beta,i] + 2[\eta\theta i][\epsilon\beta,\gamma] + 4[\eta\theta\gamma][\epsilon\beta,i]$$
$$+ 8[\eta\beta\theta][\gamma\epsilon,i] + 8[\eta\beta\theta][\gamma i,\epsilon] + 8[\eta\beta\epsilon][\gamma\theta,i] + 8[\gamma\theta\epsilon][\eta\beta,i]$$
$$+ 4[\gamma\theta i][\eta\beta,\epsilon] + 4[\eta\theta,i][\epsilon\beta,\gamma] + 4[\epsilon\beta,i][\eta\theta,\gamma]$$
$$+ 8[\eta\theta,\beta][i\gamma,\epsilon] + 8[\eta\theta,\epsilon][i\gamma,\beta] + 4[\beta\epsilon\theta][\eta,\gamma,i]\}$$
$$+ [\delta\gamma i]\{8[\beta\eta\theta][\alpha\epsilon,\zeta] + 4[\beta\eta\zeta][\alpha\epsilon,\theta] + 8[\alpha\epsilon,\theta][\beta\eta,\zeta]$$
$$+ 8[\alpha\epsilon,\zeta][\beta\eta,\theta] + 8[\alpha\zeta,\beta][\eta\epsilon,\theta]\}.$$

It should be remarked that (3.13) is an expression involving four summatory terms, and these are written as "products" merely to abbreviate.

3.6 Special cases and the covariance

There are a large number of terms in formula (3.11) for the covariance, so that it is not feasible to give details of elementary examples, apart from trivial cases. If one merely includes the totality of permutations, without regard to coincidences, there are 1137 terms for two parameters, and 12232 for three parameters. There are likely to be many similar terms, but these numbers would be the actual counts for a digitalized evaluation. In actual fact, the case of two parameters has been fully enumerated (Bowman and Shenton, K-1643) and there are 169 terms in each variance and 193 in the covariance; clearly, test cases by pencil-and-paper methods cannot be undertaken lightly.

3.6.1 Case when the parameters are equal

From (3.11), when

$$\theta_1 = \theta_2 = \ldots = \theta_h = \theta,$$

$$\mathrm{Var}_2 \hat{\theta} = -I^{-1} + I^{-3}\{2(p\Gamma_1^2\Gamma_{11}) + (p\Gamma_{1111}) + 3(p\Gamma_{11}^2) + 3(p\Gamma_1\Gamma_{111})\}$$
$$+ I^{-4}\{(p\Gamma_{111})(p\Gamma_1^3) + 7(p\Gamma_{111})^2/2 + 11(p\Gamma_{111})(p\Gamma_1\Gamma_{11})$$
$$+ 5(p\Gamma_1\Gamma_{11})^2\}. \tag{3.14a}$$

Here

$$I = \mathrm{E}\left(\frac{\partial \log p}{\partial \theta}\right)^2, \quad \Gamma_1 = \frac{\partial \log p}{\partial \theta}, \quad \Gamma_{11} = \frac{\partial^2 \log p}{\partial \theta^2}, \text{ etc.}$$

The result in (3.14a) may be written

$$\mathrm{Var}_2 \hat{\theta} = -I^{-1} + I^{-3}\left[\mathrm{E}\left(\frac{1}{p}\frac{\partial p}{\partial \theta}\right)^4 - \mathrm{E}\left(\frac{1}{p^2}\left(\frac{\partial p}{\partial \theta}\right)^2\frac{\partial^2 p}{\partial \theta^2}\right) - \mathrm{E}\left(\frac{1}{p^2}\frac{\partial p}{\partial \theta}\frac{\partial^3 p}{\partial \theta^3}\right)\right] +$$

$$+ \tfrac{1}{2}I^{-4}\left[7\left\{E\left(\frac{1}{p^2}\frac{\partial p}{\partial \theta}\frac{\partial^2 p}{\partial \theta^2}\right)\right\}^2 - 2\left\{E\left(\frac{1}{p^3}\left(\frac{\partial p}{\partial \theta}\right)^3\right)\right\}^2\right]. \quad (3.14b)$$

The above equations agree with (2.30b) and (2.33b), respectively.

3.6.2 Orthogonal parameters

These frequently, although not invariably, exist for two-parameter distributions, rarely for three-parameter distributions, and very exceptionally otherwise. They were introduced by Jeffreys (1948, p. 184) and comments on availability with respect to the number of parameters have been made by Huzurbazar (1965).

In the case of two parameters, we have $L^{12} = L^{21} = 0$, so that defining

$$a = L^{11}, \quad c = L^{22},$$

we have from (3.11)

$$\text{Cov}_2(\hat{\theta}_1, \hat{\theta}_2) = ac(aA_1^{12} + cA_2^{12}) + ac(a^2A_{11}^{12} + 2acA_{12}^{12} + c^2A_{22}^{12}), \quad (3.15a)$$

where

$$A_1^{12} = [1112] + (5[112, 1]/2 + [111, 2]/2) + 3[11, 12]$$
$$+ ([12, 1, 1] + [11, 1, 2]),$$

$$A_{11}^{12} = [111][1, 1, 2]/2 + 7[111][112]/2 + (6[112][11, 1]$$
$$+ 7[111][12, 1]/2 + 3[111][11, 2]/2) + (3[11, 1][12, 1]$$
$$+ 2[11, 1][11, 2]) + [112][1, 1, 1]/2,$$

$$2A_{12}^{12} = ([122][1, 1, 2] + [112][1, 2, 2]) + 7[112][122]$$
$$+ (15[112][12, 2]/2 + 15[122][12, 1]/2 + 7[112][22, 1]/2$$
$$+ 7[122][11, 2]/2) + (2[12, 1][22, 1] + 2[12, 2][11, 2]$$
$$+ 5[12, 1][12, 2] + [11, 2][22, 1]).$$

Similarly,

$$\text{Var}_2\hat{\theta}_1 = -a + a^2(aA_1^{11} + cA_2^{11}) + a^2(a^2A_{11}^{11} + 2acA_{12}^{11} + c^2A_{22}^{11}), \quad (3.15b)$$

where

$$A_1^{11} = [1111] + 3[111, 1] + 3[11, 11] + 2[11, 1, 1],$$
$$A_2^{11} = [1122] + (2[112, 2] + [122, 1]) + 3[12, 12] + 2[12, 1, 2],$$
$$A_{11}^{11} = [111][1, 1, 1] + 7[111]^2/2 + 11[111][11, 1] + 5[11, 1]^2,$$
$$2A_{12}^{11} = 2[112][1, 1, 2] + ([111][122] + 6[112]^2) + (13[112][12, 1] +$$

$$+ 2[111][12,2] + 6[112][11,2] + [122][11,1]) + (4[12,1]^2$$
$$+ 4[11,2][12,1] + 2[11,1][12,2]),$$
$$A_{22}^{11} = [122][1,2,2] + (5[122]^2/2 + [112][222]) + (2[122][22,1]$$
$$+ 2[112][22,2] + 6[122][12,2] + [222][12,1])$$
$$+ (2[12,2][22,1] + 2[12,1][22,2] + [12,2]^2).$$

The remaining terms in (3.15a) are derived from A_1^{12} and A_{11}^{12} by the interchange $1 \leftrightarrow 2$. It is noteworthy that (i) A_1^{11} and A_{11}^{11} are exactly the terms in Var $\hat{\theta}_1$ which arise in the single-parameter case, and (ii) the coefficients of partitions of linear terms and quadratic terms are the same – thus A_{11}^{11} has a term $7[111][112]/2$ and A_{12}^{12} has the corresponding term $\frac{1}{2}[111][122] + 3[112]^2$.

The expression corresponding to (3.15b) for $\text{Var}_2 \hat{\theta}_2$ is found by the interchange $a \leftrightarrow c$ and also the interchange $1 \leftrightarrow 2$ in every partition.

3.6.3 Two parameters with $\hat{\theta}_2$ the sample mean

Now assume that

$$\frac{\partial \log p_r}{\partial \theta_2} = (r - \theta_2) g(\theta_1, \theta_2)$$
$$= \Gamma_2^r, \tag{3.16}$$

where g is independent of r. Then $[12] = 0$ and the parameters are orthogonal. Distributions similar to this type have been studied by Sprott (1965). Writing g_1 for $\partial g/\partial \theta_1$, g_{12} for $\partial^2 g/\partial \theta_1 \partial \theta_2$ and so on, we have, from (3.15),

$$\text{Cov}_2(\hat{\theta}_1, \hat{\theta}_2) = \sum_{s=1}^{3} \lambda_s I^{-s}, \tag{3.17a}$$

$$\text{Var } \hat{\theta}_1 = \frac{1}{NI} + \frac{1}{N^2} \sum_{s=1}^{4} (b_s + B_s) I^{-s} + \ldots, \tag{3.17b}$$

where

(i) $2\lambda_1 = g_1 \mu_3 + g_{12} \mu_2^2 + 3 g_1 \mu_2 \dfrac{\partial \mu_1'}{\partial \theta_1}$,

$2\lambda_2 = 2(rp_r \Gamma_1^r \Gamma_{11}^r) + 2\mu_1'(p\Gamma_1 \Gamma_{11})$
$\qquad + (rp_r \Gamma_{111}^r) + \mu_1'(p\Gamma_{111}) - 3g_1 \mu_2 (rp_r \Gamma_1^r \Gamma_1^r),$

$\lambda_3 = -(rp_r \Gamma_1^r \Gamma_1^r)\{(p\Gamma_{111}) + 2(p\Gamma_1 \Gamma_{11})\};$

(ii) $B_1 = 0, \quad B_2 = \frac{1}{2}(g_1 \mu_2)^2 + \dfrac{\partial}{\partial \theta_1}(g_1 \mu_2)$,

$B_3 = -g_1 \mu_2 \{(p\Gamma_1^3) + 2(p\Gamma_1 \Gamma_{11})\}, \quad B_4 = 0;$

(iii) $I = (p\Gamma_1^2)$;

(iv) $b_1 = -1$, $b_3 = 3(p\Gamma_{11}^2) + 3(p\Gamma_1\Gamma_{111}) + 2(p\Gamma_1^2\Gamma_{11}) + (p\Gamma_{1111})$,

$b_2 = 0$, $b_4 = 5(p\Gamma_1\Gamma_{11})^2 + 11(p\Gamma_{111})(p\Gamma_1\Gamma_{11}) + \frac{7}{2}(p\Gamma_{111})^2 + (p\Gamma_1^3)(p\Gamma_{111})$;

(v) μ_1', μ_2, μ_3 refer to the mean, variance and third central moment of the sampled population, respectively.

3.6.4 Estimators for the negative binomial distribution

Let the probability function be

$$P_x = \frac{\alpha(\alpha+1)\ldots(\alpha+x-1)}{x!} \frac{\lambda^x \alpha^\alpha}{(\lambda+\alpha)^{\alpha+x}} \quad (x = 1, 2, \ldots) \quad (3.18)$$

$$P_0 = \left(\frac{\alpha}{\lambda+\alpha}\right)^\alpha \quad (\lambda, \alpha > 0)$$

so that $\mu_1' = \lambda$, $\mu_2 = \lambda + \lambda^2/\alpha$. Then λ and α are orthogonal parameters (as pointed out by Anscombe, 1950), and moreover, if $\theta_1 = \alpha$, $\theta_2 = \lambda$,

$$\frac{\partial \log p_x}{\partial \lambda} = (r - \lambda)g(\lambda, \alpha), \quad (3.19)$$

where $g = \alpha/\lambda(\lambda + \alpha)$. From (3.17b) the variance, in simultaneous estimation, of $\hat{\alpha}$, the maximum likelihood estimate of α, is

$$\text{Var } \hat{\alpha} = \frac{1}{NI} + \frac{1}{N^2}\sum_{s=1}^{4} \frac{k_s}{I^s} + O(N^{-3}), \quad (3.20)$$

where

(i) $I = a_{11}$, $k_1 = -1$, $k_2 = \frac{-\lambda(\lambda + 4\alpha)}{2\alpha^2(\lambda+\alpha)^2}$;

(ii) $2k_3 = a_{22} - a_{13} - 3\frac{\partial a_{12}}{\partial \alpha} + \frac{\partial^2 a_{11}}{\partial \alpha^2} - \frac{2\lambda}{\alpha(\lambda+\alpha)}\frac{\partial a_{11}}{\partial \alpha}$;

(iii) $2k_4 = -a_{12}^2 + 8a_{12}\frac{\partial a_{11}}{\partial \alpha} - 2\left(\frac{\partial a_{11}}{\partial \alpha}\right)^2$;

(iv) $a_{11} = \sum_{x=0}^{\infty} \frac{1}{P_x}\left(\frac{\partial P_x}{\partial \alpha}\right)^2$,

$a_{12} = \sum_{x=0}^{\infty} \frac{1}{P_x} \frac{\partial P_x}{\partial \alpha} \frac{\partial^2 P_x}{\partial \alpha^2}$,

$a_{22} = \sum_{x=0}^{\infty} \frac{1}{P_x}\left(\frac{\partial^2 P_x}{\partial \alpha^2}\right)^2$,

$$a_{13} = \sum_{x=0}^{\infty} \frac{1}{P_x} \frac{\partial P_x}{\partial \alpha} \frac{\partial^3 P_x}{\partial \alpha^3}.$$

The N^{-1}, N^{-2} terms in Var $\hat{\alpha}$ can now be evaluated after programming for a digital computer. Equation (3.20) has been developed from first principles from the likelihood equations, and also from the general formula (3.11), via (3.15a) and (3.15b). Bowman (1963) used (3.20) for a short tabulation over the parameter space, and Bowman and Shenton (1965, K-1643) used (3.11) for the maximum likelihood estimators of the parameters p, k corresponding to a probability generating function $(1 + p - pt)^{-k}$. Since $k = \alpha$, it is possible to see whether formula (3.11) checks out over intersecting parameter points. Some entries are given in Table 3.1.

Table 3.1
Negative binomial distribution: Var $\hat{\alpha} \sim B_1^{(2)}/N + B_2^{(2)}/N^2$

	(1,1,1)	(1,2,2)	(1,5,5)
$B_1^{(2)}$	1·216 01	6·003 00	3·218 00
$B_2^{(2)}$	8·179 02	1·849 02	5·105 01
	(2,2,1)	(2,4,2)	(2,10,5)
$B_1^{(2)}$	3·939 01	2·031 01	1·166 01
$B_2^{(2)}$	2·360 03	5·952 02	1·915 02
	(5,5,1)	(5,10,2)	(5,25,5)
$B_1^{(2)}$	2·159 02	1·165 02	7·120 01
$B_2^{(2)}$	1·223 04	3·381 03	1·207 03

(3-ples are (k, λ, p); k is index, λ is mean, $kp = \lambda$, $k = \alpha$. Two-digit entries immediately following a number, and in general throughout the rest of the monograph, refer to the power of ten by which the number is to be multiplied. It is customary to omit the sign when it is positive.)

There was agreement for the results in Table 3.1 for the two expansions used to at least 12 significant digits, indicating a satisfactory verification of the formulae.

3.7 Special cases and the bias

The number of terms in the N^{-2} bias (3.13), even with two parameters, is large. Counting all the permuted terms and ignoring coincidences, we estimate 14928 terms for two parameters, and 504522 terms for three parameters. Clearly, even if coincidences reduced these by a factor of 10 or so, it would be almost impossible to evaluate the expressions for a case with a known solution which was not at the same time trivial.

A separate development of the case when one of the ml estimators is the sample mean has been undertaken from first principles, and assuming as in (3.16) that $\partial \log p_r / \partial \theta_2 = (r - \theta_2) g(\theta_1, \theta_2)$, the bias of $\hat{\theta}_1$ turns out to be

$$E\hat{\theta}_1 = \theta_1 + \frac{1}{N}\left\{ \frac{g_1 \mu_2}{2I} - \frac{E\left(\frac{1}{p^2} \frac{\partial p}{\partial \theta_1} \frac{\partial^2 p}{\partial \theta_1^2}\right)}{2I^2} \right\}$$

$$+ \frac{1}{N^2} \sum_{s=1}^{5} \frac{(a_s^{(1)} + a_s^{(2)})}{I^s} + \ldots, \qquad (3.21)$$

where

(i) $a_1^{(1)} = 0$,

$a_2^{(1)} = -\tfrac{1}{2}(p\Gamma_{111}) - (p\Gamma_1 \Gamma_{11})$,

$8a_3^{(1)} = (p\Gamma_{11111}) + 12(p\Gamma_{11}\Gamma_{111}) + 4(p\Gamma_1\Gamma_{1111}) + 4(p\Gamma_1^2\Gamma_{111})$
$\qquad + 8(p\Gamma_1\Gamma_{11}^2)$,

$12a_4^{(1)} = 2(p\Gamma_{1111})(p\Gamma_1^3) + 18(p\Gamma_{111})(p\Gamma_1^2\Gamma_{11})$
$\qquad + 36(p\Gamma_{11}^2)(p\Gamma_1\Gamma_{11}) + 36(p\Gamma_{111})(p\Gamma_1\Gamma_{111})$
$\qquad + 36(p\Gamma_1\Gamma_{11})(p\Gamma_1\Gamma_{111}) + 36(p\Gamma_{111})(p\Gamma_{11}^2)$
$\qquad + 15(p\Gamma_{111})(p\Gamma_{111i}) + 24(p\Gamma_{1111})(p\Gamma_1\Gamma_{11})$,

$8a_5^{(1)} = 4(p\Gamma_{111})^2(p\Gamma_1^3) + 15(p\Gamma_{111})^3 + 48(p\Gamma_{111})(p\Gamma_1\Gamma_{11})^2$
$\qquad + 60(p\Gamma_{111})^2(p\Gamma_1\Gamma_{11})$;

(ii) $24a_1^{(2)} = 4g_{12}\mu_3 + 12g_1\mu_1'^2 + 12g_{12}\mu_1'\mu_2 + 3g_{122}\mu_2^2$,

$4a_2^{(2)} = 2g_1\{(r^2 p_r \Gamma_{11}^r) + 2\mu_1'(rp_r\Gamma_{11}^r)\} + 2g_{11}\frac{\partial \mu_2'}{\partial \theta_1}$
$\qquad + 2g_{12}\mu_2(rp_r\Gamma_{11}^r) + g_{111}\mu_2 + 3g_1g_{11}\mu_2^2$,

$8a_3^{(2)} = 4g_1\frac{\partial \mu_2'}{\partial \theta_1}(p\Gamma_{111}) + 2g_1\mu_2(p\Gamma_{1111}) + 3\mu_2^2 g_1^2(p\Gamma_{111})$
$\qquad + g_{11}\mu_2\{6(p\Gamma_{111}) + 8(p\Gamma_1\Gamma_{11})\} + 8g_1(rp_r\Gamma_{11}^r)^2$
$\qquad + 4g_1\mu_2(p\Gamma_{11}^2) + 4g_1\mu_2(p\Gamma_1\Gamma_{111})$,

$4a_4^{(2)} = 6g_1\mu_2(p\Gamma_{111})(p\Gamma_1\Gamma_{11}) + 3g_1\mu_2(p\Gamma_{111})^2$,

$a_5^{(2)} = 0$;

(iii) μ_1', μ_2', μ_2 are the population mean, second non-central, and third central moments respectively;

(iv) subscripts of g refer to partial derivatives with respect to θ_1, θ_2; $g_{12} = \partial^2 g/\partial \theta_1 \partial \theta_2$, for example.

As for the covariance, a comparison using the general formula (3.13) and the special formula (3.21) has been carried out for the estimation of the parameters of the negative binomial distribution (see (3.18) and (3.19); also Bowman and Shenton, 1966, ORNL-4005). There was agreement to at least 12 significant digits for overlapping points in the two parametrizations, a selection being given in Table 3.2. Further remarks on negative binomial estimation problems are deferred until Chapter 4.

Table 3.2
Negative binomial distribution: $E\hat{\alpha} \sim A_1^{(1)}/N + A_2^{(1)}/N^2$

	(1,1,1)	(1,2,2)	(1,5,5)
$A_1^{(1)}$	1·285 01	6·611 00	3·821 00
$A_2^{(1)}$	3·090 02	7·107 01	2·007 01
	(2,2,1)	(2,4,2)	(2,10,5)
$A_1^{(1)}$	2·213 01	1·215 01	7·639 00
$A_2^{(1)}$	4·526 02	1·159 02	3·829 01
	(5,5,1)	(5,10,2)	(5,25,5)
$A_1^{(1)}$	5·147 01	2·982 01	1·977 01
$A_2^{(1)}$	9·484 02	2·665 02	9·763 01

(3-ples (k, λ, p): k is index, λ is mean, $kp = \lambda$, $k = \alpha$.)

3.8 Adjusted order of magnitude method

For the single-parameter case (compare **2.6**) this method was quite effective. In the present situation, we should write (3.1) as

$$(n\Gamma_\alpha) + x_r\{(p\Gamma_{r\alpha}) + (\epsilon\Gamma_{r\alpha})\} + \tfrac{1}{2}x_s x_t(n\Gamma_{st\alpha}) + \ldots = 0 \qquad (3.22a)$$

and formally multiply throughout by $L^{\alpha\lambda}$ to give

$$(n\Gamma_\alpha)L^{\alpha\lambda} + x_\lambda + x_r(\epsilon\Gamma_{r\beta})L^{\alpha\lambda} + \tfrac{1}{2}x_s x_t L^{\alpha\lambda}(n\Gamma_{st\alpha}) + \ldots = 0. \quad (3.22b)$$

Construct a similar expression for x_μ, and then first-order terms in expectation could be extracted from $x_\lambda x_\mu$. The approach clearly runs into complications for N^{-2} terms.

REFERENCES

1. Anscombe, F. J. (1950). "Sampling theory of the negative binomial and logarithmic series distributions". *Biometrika*, 37, 358–82.
2. Bowman, K. O. (1963). "Moments to higher orders for maximum likelihood estimators, with an application to the negative binomial distribution". Thesis in support of the Ph.D. degree at Virginia Polytechnic Institute.

3. Bowman, K. O. and Shenton, L. R. (1965). "Asymptotic covariances for the maximum likelihood estimators of the parameters of a negative binomial distribution". Report K-1643, Union Carbide Nuclear Division, Oak Ridge, Tenn.
4. Bowman, K. O. and Shenton, L. R. (1966). "Biases of estimators for the negative binomial distribution". Report ORNL-4005, Oak Ridge National Lab., Oak Ridge, Tenn.
5. Good, I. J. (1959). "Generalizations to several variables of Lagrange's expansion, with applications to stochastic processes". *Proc. Camb. Phil. Soc.*, **56**, 367–80.
6. Haldane, J. B. S. (1953). "The estimation of two parameters from a sample". *Sankhyā*, **12**, 313–20.
7. Huzurbazar, V. S. (1965). "Sufficient statistics and orthogonal parameters". *Sankhyā*, **17**, 217–20.
8. Jeffreys, H. (1948). *Theory of Probability* (3rd edn, 1961). Oxford, at the Clarendon Press.
9. Shenton, L. R. and Wallington, P. A. (1962). "The bias of moment estimators with an application to the negative binomial distribution". *Biometrika*, **49**, 193–204.
10. Sprott, D. A. (1965). "A class of contagious distributions and maximum likelihood estimation". In *Classical and Contagious Discrete Distributions Symposium*, ed. G. P. Patil, Statistical Publishing Company, Calcutta (Pergamon Press, Oxford), 337–50.

4 BIASES AND COVARIANCES FOR ESTIMATORS IN NON-REGULAR CASES[(*)]

4.1 Introduction

In section **1.3** we attempted to classify estimators from the viewpoint of numerical tractability. Under *1.3.4*(c) we mentioned estimators which are implicit functions of the sample, and for which the likelihood function can only be expanded as a Taylor series in n dimensions. Attempts to reduce the dimensionality by what we have called the "adjusted order of magnitude method" run into serious complications. The present chapter deals with those "non-regular" cases for which there seems to be no other recourse but to use the full-length formulae for biases, etc., as far as the N^{-2} terms developed in Chapter 3.

The application of formulae (3.11) for the N^{-2} covariances and (3.13) for N^{-2} bias requires the facilities of a fairly large digital computer. The fundamental entities are logarithmic derivatives, all fifth order and lower for biases, fourth order and lower for covariances. If a discrete variate is involved, the probability function may be given implicitly only (N.T.A., for example), in which case parameter derivatives will also be available implicitly only.

Care and constant attention are needed to detect loss of accuracy such as may arise from massive calculations involving small- and large-number sums and products, with a risk of overflows or underflows. Whenever simple checks can be derived, they should be implemented. For example, if the product of estimators turns out to be a sample moment, then equality of N^{-1}, N^{-2} coefficients provides a check. Smoothness, or the absence of discontinuities, in tabulations over equally spaced lattices in the parameter space provides a crude acceptance criterion.

Since about 15000 terms may arise in N^{-2} terms of biases for two parameters, half-a-million for three parameters, and perhaps 25 million for four parameters (even if these are 10 to 50 times too large), it is obvious that in such cases there is no possibility of evaluation by elementary computing approaches. Moreover, four- or five-parameter situations (mixtures of normal distributions, gamma distributions, although ml estimation is not recommended in these cases) may mark limiting cases for present facilities.

[(*)] Estimation of the parameters of: (a) the negative binomial, (b) Neyman Type A, (c) Pólya–Aeppli, (d) three-parameter gamma, (e) two-parameter Hermite distribution, and an exponential regression model are considered.

Beyond these, the realm of trans-scientific structures is reached, and the dubious returns pall in comparison to the outlay of effort and facilities demanded.

In the remainder of this chapter, several examples of N^{-1} and N^{-2} biases and covariances are set out. A minimum of formulae is given, and the reader is advised to study the numerical evaluations to gain an insight into the subject. Parameter points have been chosen from large tabulations to demarcate dangerous regions and facilitate comparisons of parameter estimators. It is impossible to be certain that generalizations, when they are made, are strictly correct. The aim is to sketch out approximately, and economically from the point of view of the space available, boundaries of regions in the parameter space where N^{-2} terms may or may not be ignored.

4.2 Negative binomial distribution (N.B.D.)

pgf: $(1 + p - pt)^{-k}$, $p, k > 0$.

probability function: $q^{-k} \dfrac{\Gamma(k+x)}{x!\,\Gamma(k)} \left(\dfrac{p}{q}\right)^x$, $(x = 0, 1, \ldots; q = p + 1)$.

moments: mean $= kp$, variance $= kpq$.

estimators: ml

$$\frac{n_1}{\hat{k}} + n_2\left(\frac{1}{\hat{k}} + \frac{1}{\hat{k}+1}\right) + \ldots + n_s\left(\frac{1}{\hat{k}} + \ldots + \frac{1}{\hat{k}+s-1}\right) = \ln\left(1 + \frac{m_1}{\hat{k}}\right),$$

$$\hat{p}\hat{k} = m_1. \qquad (4.1)$$

moments

$$k^* = m_1^2/(m_2 - m_1) \quad p^* = (m_2 - m_1)/m_1.$$

Table 4.1: Bias and covariances

COMMENTS

(a) It is suggested from the form of the moment estimators that p^* may be better behaved than k^* from the viewpoint of stability of moment series; k^* has a singularity in the denominator of a serious form, involving a non-negligible region of the sample space, whereas p^* can only erupt when $m_1 = 0$, with a small probability in general. From extensive tabulations of biases and covariances, it can be concluded that $\hat{k}$ and k^* have far less stable moment series than $\hat{p}, p^*$.

(b) There are regions in the parameter space for which the first-order term in the bias or covariance can be grossly or partly misleading. For the variance of $\hat{k}$ or k^*, samples larger than 100 are required to damp-off the second-order term to $\frac{1}{4}$ the first-order term for $p \leqslant 2$, $k \leqslant 190$; this region is enlarged as k decreases and care should be exercised whenever $k \leqslant 0.1$, $p \leqslant 9$; $0.1 <$

Table 4.1

Bias and covariances for negative binomial distributions

	$k=p=0.1$		$k=0.1 \quad p=1$		$k=0.1 \quad p=0.5$		$k=0.1 \quad p=5$	
E $\hat{k}$	2·48—00	51·0	5·37—02	1·02	1·47—01	2·89	1·04—02	0·17
E k^*	2·78—00	44·8	1·10—01	−0·51	2·31—01	0·83	4·49—02	−0·90
E $\hat{p}$	3·15—02	0·60	4·59—02	0·09	3·83—02	0·17	9·37—02	0·03
E p^*	−1·21—01	−0·10	−2·20—01	0·10	−1·65—01	−0·10	−6·60—01	0·10
Var $\hat{k}$	2·51—01	146·7	5·83—03	2·50	1·55—02	7·64	1·22—03	0·35
Var k^*	2·66—01	143·0	8·80—03	0·45	1·98—02	5·41	3·17—03	−1·56
Cov $(\hat{k},\hat{p})$	−2·51—01	50·7	−5·83—02	0·99	−7·73—02	2·83	−6·50—02	0·14
Cov (k^*,p^*)	−2·66—01	45·6	−8·80—02	−0·89	−9·39—02	0·01	−1·58—01	−1·42
Var $\hat{p}$	2·62—01	1·79	7·83—01	0·27	4·61—01	0·44	6·06—00	0·11
Var p^*	2·77—01	−1·55	1·08—00	−0·62	5·70—01	−0·73	1·09 01	−0·52
	$k=5.0 \quad p=0.1$		$k=0.1 \quad p=8.0$		$k=5.0 \quad p=3.0$		$k=50.0 \quad p=2.0$	
E $\hat{k}$	1·47 01	7·72	7·58—03	0·12	2·40—01	0·07	3·00—00	0·09
E k^*	1·52 01	7·60	4·02—02	−0·91	2·93—01	0·05	3·06—00	0·08
E $\hat{p}$	−8·32—03	0·17	1·23—01	0·17	−3·57—02	0·00	−2·97—00	0·00
E p^*	−1·32—02	0·00	−9·90—01	−0·10	−4·80—01	0·00	−3·06—02	0·00
Var $\hat{k}$	7·13 01	21·3	9·11—04	0·23	9·00—01	0·22	1·13 02	0·29
Var k^*	7·26 01	21·4	2·78—03	−1·66	1·07—00	0·19	1·15 02	0·29
Cov $(\hat{k},\hat{p})$	−1·43—00	7·85	−7·29—02	0·12	−5·40—01	0·07	−4·51—00	0·10
Cov (k^*,p^*)	−1·45—00	7·80	−2·23—01	−1·44	−6·40—01	0·04	−4·59—00	0·09
Var $\hat{p}$	2·88—02	0·08	1·30 01	0·09	3·48—01	−0·06	1·82—01	−0·10
Var p^*	2·93—02	−0·04	2·50 01	−0·51	4·08—01	−0·23	1·85—01	−0·11

(A typical term being $A_1/N + A_2/N^2$ or $(A_1/N)\{1 + A_2/(NA_1)\}$, the entries are A_1/N and $A_2/(NA_1)$ when $N=100$. If the second term is less than unity, a sample of $N=100$ is adequate to damp off the second-order terms; if it is large, then $N=100$ is not adequate and a simple scaling factor can be worked out.)

$k \leqslant 0.2$, $p \leqslant 5$; $0.2 < k \leqslant 0.5$, $p \leqslant 4$; $0.5 < k \leqslant 0.6$, $p \leqslant 3$. Very approximately, the "bad" region is $p > 0$, $k > 0$, $kp^2 < 6$, which approximately matches the condition for the N.B.D. to be tangential to the Poisson. The worst case evaluated is $k = p = 0.1$, when a sample of a million would be required to put reliance on the usual asymptotic for Var $\hat{k}$.

For Cov $(\hat{k},\hat{p})$, large samples are needed for p and k small, and in particular N must exceed 100 for the region $k \leqslant 0.1$, $p \leqslant 4$; $0.1 < k \leqslant 0.2$, $p < 2$; $0.2 < k \leqslant 1.0$, $p \leqslant 1$.

For Var $\hat{p}$, the bad region is $k \leqslant 0.1$, $p \leqslant 1.0$; $0.1 < k \leqslant 0.2$, $p \leqslant 0.6$; $0.2 < k \leqslant 0.4$, $p \leqslant 0.2$; $0.4 < k < 1.0$, $p \leqslant 0.1$. For an approximate descriptive, the troublesome region is $p > 0$, $k > 0$, $p^2 k < 0.1$.

For the biases, E $\hat{k}$ is similar to Var $\hat{k}$, with a slightly less region of large-sample requirements; also E $\hat{p}$ is similar to Var $\hat{p}$ but rather better behaved (see Fig. 4.1 to 4.8).

(c) From the evaluations, there is evidence that when the asymptotic efficiency is high (either that of p^*, k^* or joint estimators), second-order terms are important, and when asymptotic efficiency is low, second-order terms can be ignored. Put otherwise, high asymptotic efficiency is not efficient at all unless there is compensation in the direction of large or very large samples.

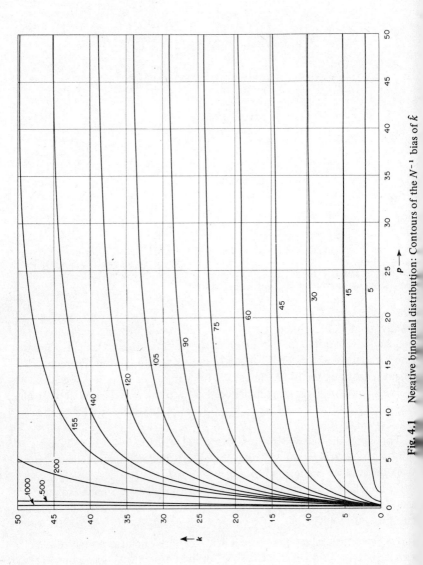

Fig. 4.1 Negative binomial distribution: Contours of the N^{-1} bias of $\hat{k}$

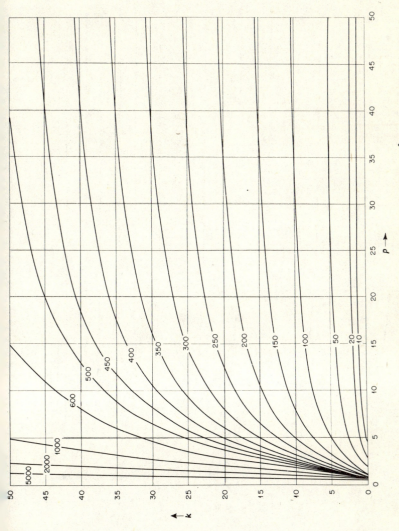

Fig. 4.2 N.B.D.: Contours of the N^{-2} bias of $\hat{k}$

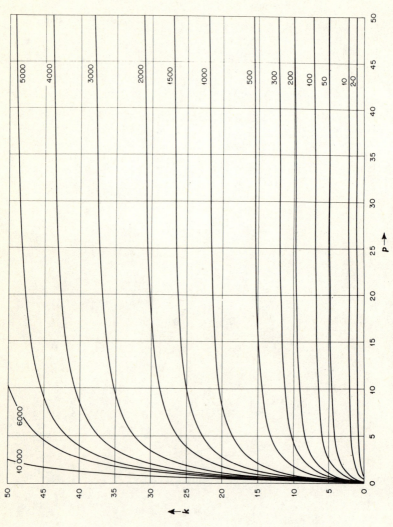

Fig. 4.3 N.B.D.: Contours of the N^{-1} variance of $\hat{k}$

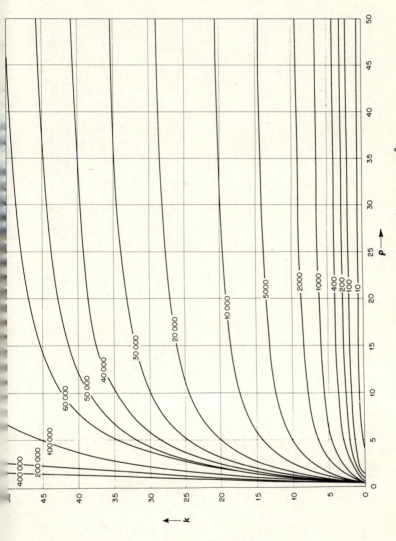

Fig. 4.4 N.B.D.: Contours of the N^{-2} variance of $\hat{k}$

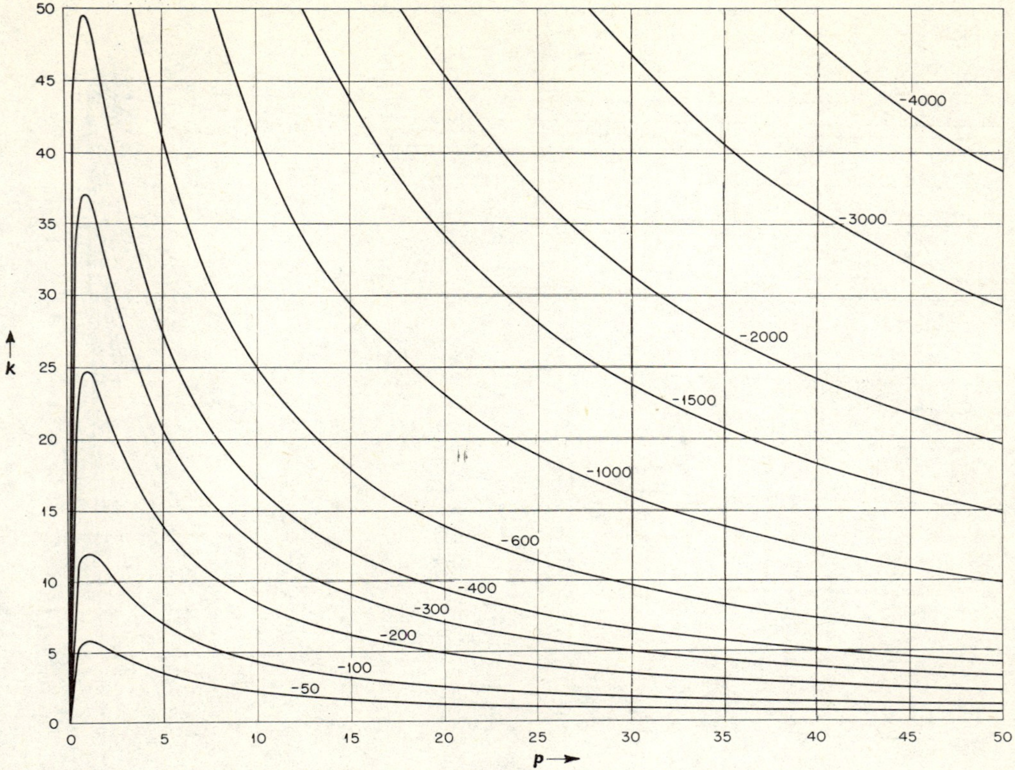

Fig. 4.5. NBD: Contours of the N^{-1} covariance of $\hat{k}, \hat{p}$

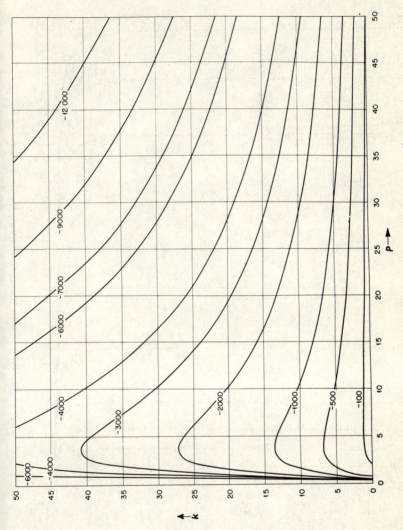

Fig. 4.6 N.B.D.: Contours of the N^{-2} covariance of $\hat{k}, \hat{p}$

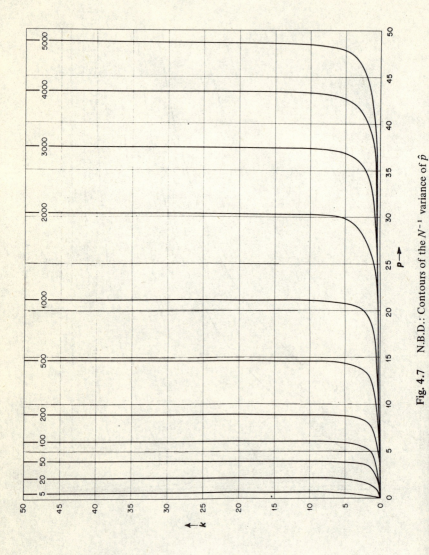

Fig. 4.7 N.B.D.: Contours of the N^{-1} variance of $\hat{p}$

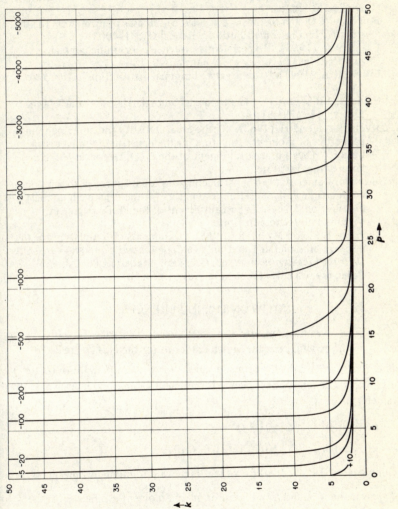

Fig. 4.8 N.B.D.: Contours of the N^{-2} variance of $\hat{p}$

REFERENCES

1. Anscombe, F.J. (1950). "Sampling theory of the negative binomial and logarithmic series distributions". *Biometrika*, **37**, 358–82.
2. Bowman, K.O. and Shenton, L.R. (1965). "Asymptotic covariances for the maximum likelihood estimators of the parameters for a negative binomial distribution". Report K-1643, Union Carbide Corporation Nuclear Division.
3. Bowman, K.O. and Shenton, L.R. (1966). "Biases of estimators for the negative binomial distribution". Report ORNL-4005.
4. Evans, D.A. (1953). "Experimental evidence concerning contagious distributions in ecology". *Biometrika*, **40**, 186–211.
5. Fisher, R.A. (1941). "The negative binomial distribution". *Ann. Eugen.*, **11**, 182–7.
6. Haldane, J.B.S. (1941). "The fitting of binomial distributions". *Ann. Eugen.*, **11**, 179–81.
7. Myers, Raymond H. (1963). "Orthogonal statistics and some sampling properties of moment estimators for the negative binomial distribution." Thesis in part fulfilment of the Ph.D. Degree at Virginia Polytechnic Institute.
8. Sheehan, Daniel M. (1967). "The computational approach to sampling moments: a study of certain estimators for the negative binomial distribution." Thesis in part fulfilment of the Ph.D. Degree at Virginia Polytechnic Institute.
9. Shenton, L.R. and Myers, R. (1965). "Comments on estimation for the negative binomial distribution". In *Classical and Contagious Discrete Distributions Symposium*, ed. G.P. Patil, Statistical Publishing Company, Calcutta (Pergamon Press, Oxford).

NOTES ON THE REFERENCES

(2) This gives the complete analytical form for $\text{Var}_2\,\hat{\theta}_1$, $\text{Cov}_2(\hat{\theta}_1, \hat{\theta}_2)$, $\text{Var}_2\,\hat{\theta}_2$ in the general case; the individual terms for the N.B.D. are listed.

Tabulations N^{-1} and N^{-2} terms in $\text{Var}\,\hat{k}$, $\text{Var}\,k^*$, $\text{Cov}(\hat{k}, \hat{p})$, $\text{Cov}(k^*, p^*)$; $\text{Var}\,\hat{p}$, $\text{Var}\,p^*$; generalized efficiency for moment estimators. In each case the parameter space is

$$p = 0{\cdot}1(0{\cdot}1)1{\cdot}0(1{\cdot}0)10(5)50$$

$$k = 0{\cdot}1(0{\cdot}1)1{\cdot}0(1{\cdot}0)12,\ 15(5)70,\ (10)190,$$

excluding $p \geqslant 10$ and $k \geqslant 80$ for which there were rounding-off problems.

(3) gives the N^{-1} and N^{-2} biases for ml and moment estimators for the parameter space

$$p = 0{\cdot}1(0{\cdot}1)2{\cdot}4(0{\cdot}2)5{\cdot}4,\ 6(0{\cdot}5)15(1)23,\ 25(5)50$$

$$k = 0{\cdot}1(0{\cdot}1)2{\cdot}4(0{\cdot}2)7{\cdot}2,\ 7{\cdot}5(0{\cdot}5)19(1)31;\ 35(5)60.$$

(There is one page out of order: p. 41 refers to $\hat{k}$.)

This reference also considers numerical accuracy of the tabulations and gives a tabulation of identities for N^{-1} and N^{-2} terms in

$$\text{cov}(\hat{p}, \hat{k}) = pk - (E\,\hat{p})(E\,\hat{k}).$$

(8) gives a comprehensive study of several estimators, including moments and ml estimators, mixed moment and frequency estimators, and estimators based on linear functions of observed frequencies. In addition there is work on the admissibility of estimators, and results of extensive simulation runs on their distributional properties.

4.3 Neyman Type A distribution (N.T.A.D.)

pgf: $e^{\theta_1(e^{\theta_2(t-1)}-1)}$ $\quad \theta_1, \theta_2 > 0$

probability function: $P_x = \dfrac{e^{-\theta_1}\theta_2^x}{x!}\left(0^x + \dfrac{\lambda 1^x}{1!} + \dfrac{\lambda^2 2^x}{2!} + \ldots\right),$ (4.2)

where $\lambda = \theta_1 e^{-\theta_2}$, $0^x = 1, x = 0$
$\qquad\qquad\qquad = 0, x \neq 0,$

moments: mean $= \theta_1\theta_2$, variance $= \theta_1\theta_2(1+\theta_2)$.

recurrence: $\begin{cases} P_{x+1} = \dfrac{\theta_1\theta_2 e^{-\theta_2}}{x+1}\sum_{s=0}^{x}\dfrac{\theta_2^s}{s!}P_{x-s}, & (x = 0, 1, \ldots); \\ P_0 = e^{\lambda-\theta_1}. \end{cases}$ (4.3)

estimators: ml

$$\frac{1}{N}\sum' n_x(x+1)\frac{\hat{P}_{x+1}}{\hat{P}_x} = \hat{\theta}_1\hat{\theta}_2$$

$$= m_1,$$

where (i) n_x is the frequency of the variate x,
(ii) $\hat{P}_x = P_x$ when $\theta_1 = \hat{\theta}_1, \theta_2 = \hat{\theta}_2$,
(iii) Σ' is summation over the sample.

(Moment estimators as in 4.2.)

Table 4.2: Bias and covariances

COMMENTS

(a) The parameter regions of poor performance for second-order terms are similar to those for the N.B.D. (θ_1, θ_2 corresponding to k, p respectively. The second-order terms for $E\,\hat{\theta}_1$, $\text{Var}\,\hat{\theta}_1$, $\text{Cov}(\hat{\theta}_1, \hat{\theta}_2)$ are important, especially

Table 4.2

Bias and covariances for Neyman Type A distribution: evaluations for $N = 100$

	$\theta_1 = 0.5$ $\theta_2 = 0.5$	$\theta_1 = 25$ $\theta_2 = 0.5$	$\theta_1 = 5$ $\theta_2 = 15$	$\theta_1 = 5$ $\theta_2 = 25$	$\theta_1 = 15$ $\theta_2 = 25$
(a)	·250 + ·163	·206 + ·089	·02488 + ·01835	·00589 + ·00611	·03335 −·03881
(a*)	·410 + ·194	·214 + ·092	·03358 + ·00103	·03212 + ·00092	·03206 + ·00102
(b)	−·0735 −·0316	−·1511 −·0043	−·0756 −·2085	−·01842 −·10967	−·06330 + 1·18492
(b*)	−·3500 −·0100	−·1540 + ·0002	−·0540 + ·0000	−·05240 + ·00002	−·05213 + ·00000
(c)	·2500 + ·3947	·1765 + ·2296	·01178 + ·01108	·00421 + ·00185	·02136 + ·01031
(c*)	·3600 + ·5736	·1835 + ·2398	·02477 + ·00284	·02364 + ·00255	·02230 + ·00260
(d)	−·2550 −·1580	−·1760 −·0829	−·00977 −·00963	−·00221 −·001698	−·02069 −·008165
(d*)	−·3400 −·1650	−·1832 −·0853	−·02277 −·00067	−·02164 −·000587	−·02163 −·000688
(e)	·2750 + ·01247	·1768 −·00184	·00991 + ·00856	·00229 + ·001602	·02072 + ·006943
(e*)	·3800 −·03180	·1840 −·00229	·02291 −·00029	·02172 −·000270	·02166 −·000234

((a) through (e) refer to E $(\hat{\theta}_1 - \theta_1)/\theta_1$, E $(\hat{\theta}_2 - \theta_2)/\theta_2$, Var $(\hat{\theta}_1/\theta_1)$, Cov $(\hat{\theta}_1/\theta_1, \hat{\theta}_2/\theta_2)$, Var $(\hat{\theta}_2/\theta_2)$, respectively; (a*) through (e*) refer to corresponding moments for moment estimators. The two entries are N^{-1} and N^{-2} terms; thus, under (a) of $\theta_1 = 0.5$, $\theta_2 = 0.5$, ·250 refers to the N^{-1} term and ·163 to the N^{-2} term.)

for θ_2 small and θ_1 large; as before, the series for E θ_2 and Var $\hat{\theta}_2$ are relatively benign (see Fig. 4.9 to 4.18). However, there is now a new feature, for second-order terms become very prominent in a wedge-shaped region roughly following the line $3\theta_1 = 2\theta_2$. What is the explanation for this? It could be related to the fact that the N.T.A.D. is multimodal, and can have, as pointed out by Anscombe (1950), an indefinitely large number of modes, the modes being spaced at multiples of θ_2 in the region of high modality (Fig. 4.19). The multimodality question has also been studied by Barton (1957), who gives a diagram of the double points of the bounding arcs corresponding to a specified modality.

(b) Large to very large samples are required to damp out N^{-2} terms, especially for E $\hat{\theta}_1$ and Var $\hat{\theta}_1$ in a thin strip for which θ_2 is small, and also in a region roughly defined by the triangle with vertices $(\theta_1, \theta_2) = (5, 10)$, $(1, 25)$, $(15, 25)$. The dangerous regions are graphically described in Fig. 4.20 and 4.21.

(c) As with the N.B.D., high asymptotic efficiency of moment methods tends to occur in the region of low stability of moment series, and low asymptotic efficiency corresponds to stable moment series (see Fig. 4.22). For example, if the ratio of the covariance determinants $|\text{Cov}(\hat{\theta}_i, \hat{\theta}_j)| \div |\text{Cov}(\theta_i^*, \theta_j^*)|$ is expressed as $E_f (1 + \hat{c}_1/N) \div (1 + c_1^*/N)$, then at $\theta_1 = 1$, $\theta_2 = 20$, $E_f = 0.056$, $\hat{c}_1 = 0.07$, $c_1^* = 0.16$; but at $\theta_1 = 25$, $\theta_2 = 0.5$, $E_f = 0.961$, $\hat{c}_1 = 52$, $c_1^* = 57$. Interestingly enough, at $\theta_1 = 1$, $\theta_2 = 20$ there are modes at $x = 0, 20, 40$.

REFERENCES

1. Anscombe, F.J. (1950). "Sampling theory of the negative binomial and logarithmic series distributions". *Biometrika*, 37, 358.
2. Barton, D.E. (1957). "The modality of Neyman's contagious distribution of Type A". *Trabajos Estadist.*, 8, 15.
3. Bowman, K.O. and Shenton, L.R. (1967). "Remarks on estimation problems for the parameters of the Neyman Type A distribution". Report ORNL-4102.
4. Shenton, L.R. and Bowman, K.O. (1967). "Remarks on large sample estimators for some discrete distributions". *Technometrics*, 9, 4, pp. 587–98.

NOTES ON THE REFERENCES

(3) contains the tabulations:

(i) Bias of θ_1^* (moment estimator), N^{-1} to N^{-5} terms

$$\theta_1 = 1(1)25, \quad \theta_2 = 1(1)25.$$

(ii) Bias of $\hat{\theta}_1$ and θ_1^*, N^{-1} and N^{-2} terms. *[To page 108, below]*

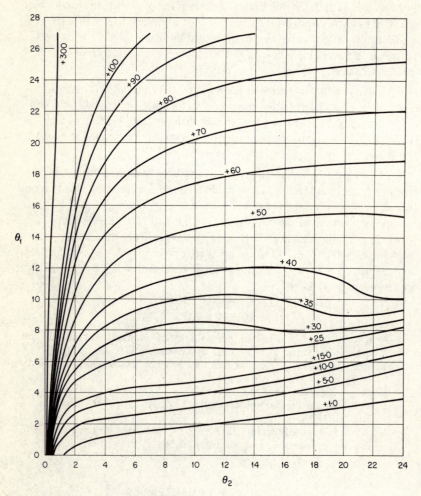

Fig. 4.9 Neyman Type A distribution: First-order term in bias of $\hat{\theta}_1$

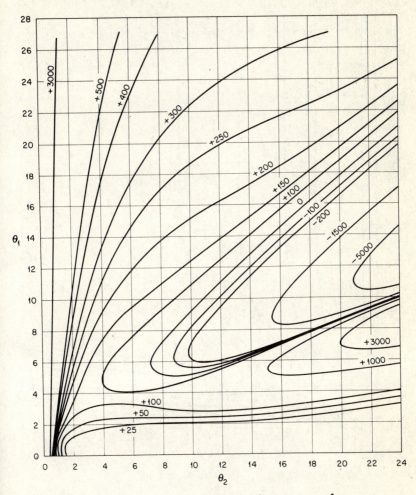

Fig. 4.10 N.T.A.D.: Second-order term in bias of $\hat{\theta}_1$

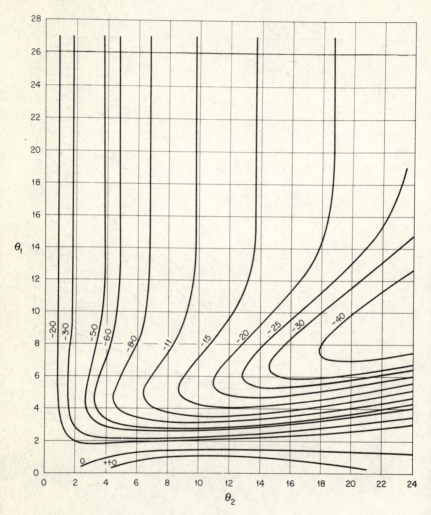

Fig. 4.11 N.T.A.D.: First-order term in bias of $\hat{\theta}_2$

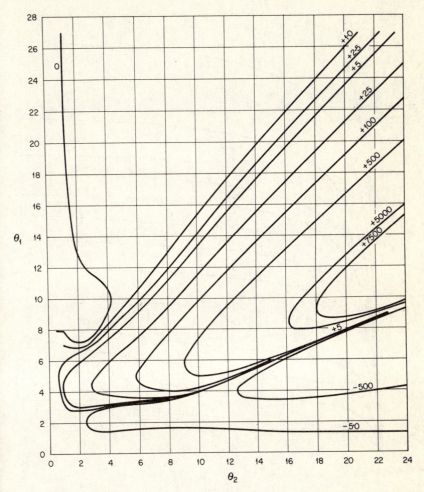

Fig. 4.12 N.T.A.D.: Second-order term in bias of $\hat{\theta}_2$

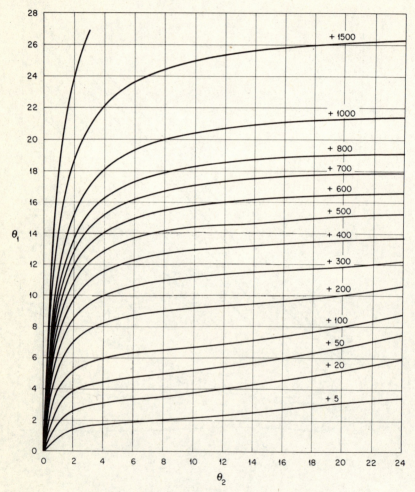

Fig. 4.13 N.T.A.D.: First-order term in variance of $\hat{\theta}_1$

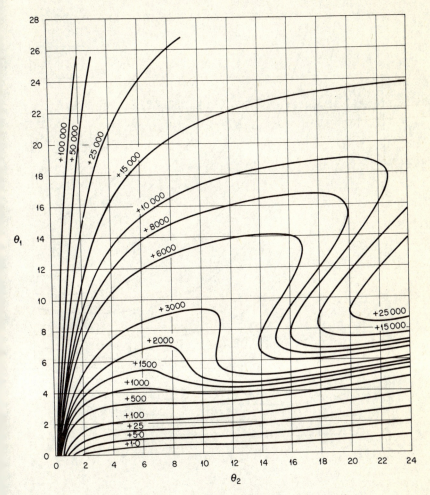

Fig. 4.14 N.T.A.D.: Second-order term in variance of $\hat{\theta}_1$

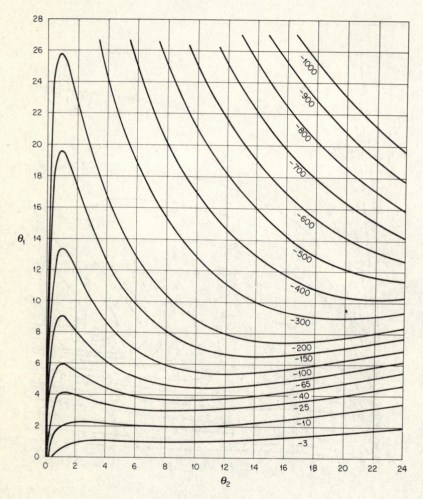

Fig. 4.15 N.T.A.D.: First-order term in Cov $(\hat{\theta}_1, \hat{\theta}_2)$

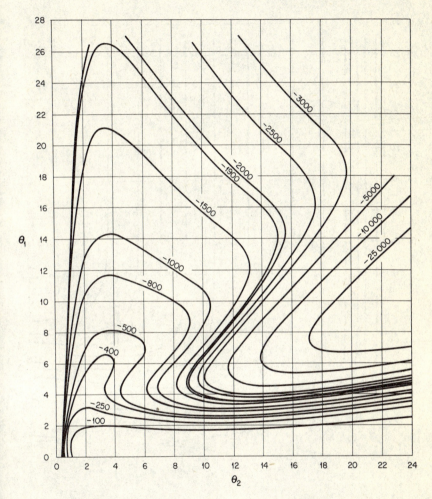

Fig. 4.16 N.T.A.D.: Second-order term in Cov $(\hat{\theta}_1, \hat{\theta}_2)$

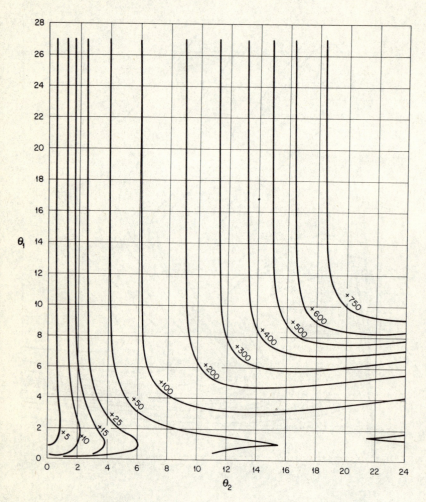

Fig. 4.17 N.T.A.D.: First-order term in variance of $\hat{\theta}_2$

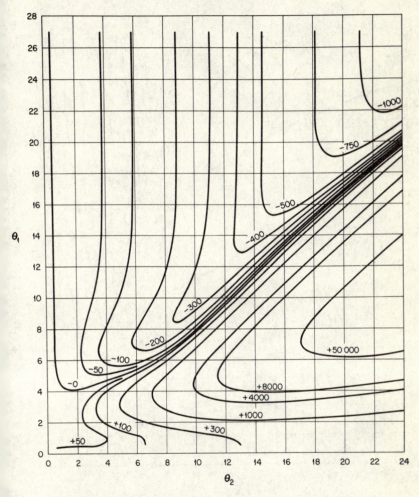

Fig. 4.18 N.T.A.D.: Second-order term in variance of $\hat{\theta}_2$

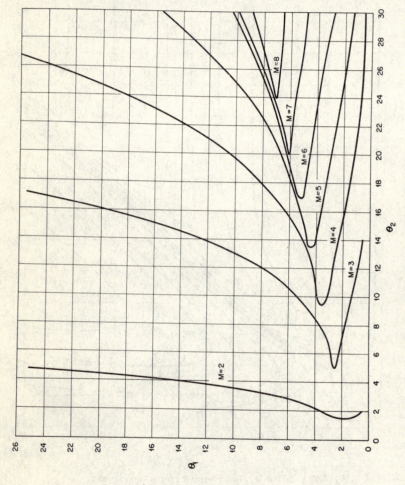

Fig. 4.19 N.T.A.D.: Number of modes

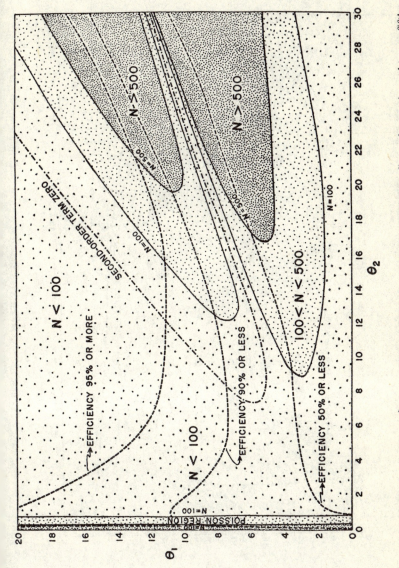

Fig. 4.20 N.T.A.D.: Bias of $\hat{\theta}_1$: sample size to make ratio (second-order term to first-order term) equal to one-fifth

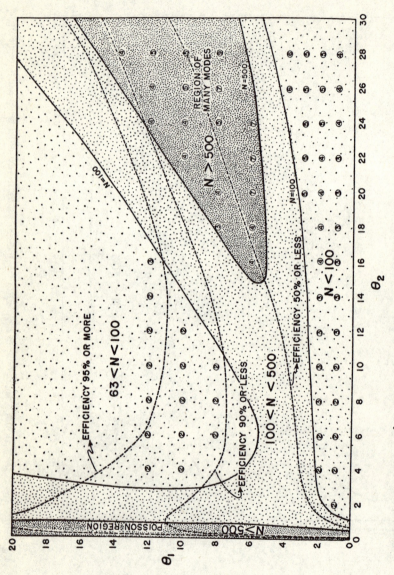

Fig. 4.21 N.T.A.D.: Var $\hat{\theta}_1$: sample size to make ratio (second-order term to first-order term) equal to one-fifth

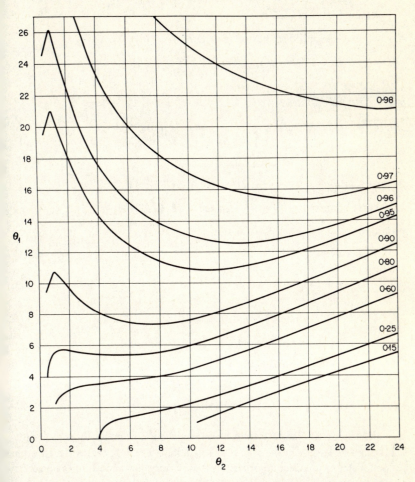

Fig. 4.22 N.T.A.D.: Asymptotic efficiency of the method of moments (first-order terms only)

(iii) Bias of $\hat{\theta}_2$ and θ_2^*, N^{-1} and N^{-2} terms.
(iv) Cov$(\hat{\theta}_i, \hat{\theta}_j)$ and Cov(θ_i^*, θ_j^*), $i = j = 1, 2$.
(v) Dominant terms in asymptotic efficiency in joint estimation of θ_1, θ_2 by moments.
(In (ii)–(iv) the parameter space is $\theta_1, \theta_2 = 0.5, 1, 5, 10(5)25$.)

4.4 Pólya–Aeppli distribution (P.A.D.)

pgf: $\exp s(t-1)/(1-t\zeta)$, $0 < \zeta < 1$
$s > 0$.

probability function:

$$xP_x = (2\zeta(x-1) + s(1-\zeta))P_{x-1} - \zeta^2(x-2)P_{x-2}, \quad x = 2, 3, \ldots$$
$$P_0 = e^{-s}, \quad P_1 = s(1-\zeta)P_0.$$
(4.4)

parametrization: $\theta_1 = s/(2\zeta)$, $\theta_2 = 2\zeta/(1-\zeta)$.

moments: mean $= \theta_1\theta_2$, variance $= \theta_1\theta_2(1+\theta_2)$.

estimation: ml

$$\hat{\theta}_1\hat{\theta}_2 = m_1$$
$$\sum' n_x(x+1)\frac{\hat{P}_{x+1}}{\hat{P}_x} = Nm_1$$
(4.5)

($\hat{P}_x = P_x$ when $\theta_1 = \hat{\theta}_1, \theta_2 = \hat{\theta}_2$).

(Moment estimators as in **4.2**.)

Table 4.3: Bias and covariances

COMMENTS

(a) There is a similar pattern of behaviour for the moments here as for the N.B.D. and N.T.A.D. In the region of the parameter space tangential to the Poisson distribution, second-order terms are important and samples in excess of 100 are required for stability. In addition, θ_2 small is usually associated with lack of stability for the moments of $\hat{\theta}_1$, including Cov$(\hat{\theta}_1, \hat{\theta}_2)$.

(b) The *correlation* between second-order terms for ml and moment estimators is noteworthy, and this characteristic is quite marked for θ_1 estimators.

(c) The inverse "efficiency law" (see section **4.3**(c)) is supported by this case. Examples of E_f, $\hat{c}_1$, c_1^* are as in the table on page 110.

Table 4.3

Bias and covariances for the Pólya–Aeppli distribution: evaluations for $N = 100$

	$\theta_1 = \theta_2 = 0.5$	$\theta_1 = 25$	$\theta_2 = 0.5$	$\theta_1 = 0.5$	$\theta_2 = 1.0$	$\theta_1 = 1.0$	$\theta_2 = 25$	$\theta_1 = 20$	$\theta_2 = 15$
(a)	·3400 + ·2628	·2093	+ ·0916	·1146	+ ·0290	·0171	+ ·0006	·0329	+ ·0001
(a*)	·5100 + ·2694	·2160	+ ·0932	·2200	+ ·0204	·0432	− ·0006	·0340	+ ·0011
(b)	− ·0070 − ·0010	− ·0292	− ·0001	− ·0048	− ·0002	− ·0063	− ·0000	− ·0107	+ ·0000
(b*)	− ·0800 + ·0005	− ·0310	+ ·0000	− ·0500	+ ·0003	− ·0158	+ ·0001	− ·0110	+ ·0000
(c)	·3432 + ·6836	·1803	+ ·2367	·1200	+ ·0716	·0157	+ ·0013	·0224	+ ·0029
(c*)	·4400 + ·8436	·1852	+ ·2436	·1800	+ ·0908	·0325	+ ·0009	·0233	+ ·0029
(d)	− ·3328 − ·2592	− ·1801	− ·0854	− ·1100	− ·0284	− ·0108	− ·0005	− ·0222	− ·0008
(d*)	− ·4300 − ·2292	− ·1850	− ·0866	− ·1700	− ·0097	− ·0275	+ ·0012	− ·0231	− ·0008
(e)	·3823 + ·0230	·1812	− ·0014	·1400	+ ·0050	·0162	+ ·0001	·0224	− ·0002
(e*)	·4800 − ·0561	·1860	− ·0024	·2000	− ·0187	·0329	− ·0013	·0234	− ·0003

(Key: See footnote to Table 4.2).

θ_1	θ_2	E_f	$\hat{c}_1$	c_1^*
25	0·5	0·97	55	58
5	0·5	0·90	12	15
0·5	15	0·38	·06	·09

Thus, high asymptotic efficiency seems to imply dominance of second-order terms, with the requirement of great care over sample sizes.

REFERENCES

1. Shenton, L.R. and Bowman, K.O. (1967). "Sampling moments of moment and maximum likelihood estimators for discrete distributions" (appendix by Juris Reinfelds). Presented at 36th Session of International Statist. Inst., Sydney, Australia.
2. Shenton, L.R. and Bowman, K.O. (1967). "Remarks on large sample estimators for some discrete distributions". *Technometrics*, 9, 4, pp. 587–98.

NOTES ON THE REFERENCES

(1) gives suggestions as to how to set up digital programs for the derivatives of probability functions given recursively. The method has been used on the N.T.A.D.

The bias of θ_1^*, Var θ_1^*, Cov (θ_1^*, θ_2^*), for the moment estimators are given to order N^{-6} for the parameter space $\theta_1, \theta_2 = 0.5, 1.0, 5(5)25$, the entries being evaluated for $N = 100$.

In addition, for the same parameter space, the N^{-1}, N^{-2} biases and covariances of ml and the moment estimators are given.

4.5 Three-parameter Gamma distribution

density:
$$f(x; m, a, \rho) = X^{\rho-1} e^{-X}/a\, \Gamma(\rho), \quad X > 0 \\ = 0, \quad X < 0; \tag{4.6}$$

$$X = (x - m)/a, \quad a, \rho > 0, m \text{ arbitrary real.}$$

moments:
$\mu_1' = m + a\rho$
$\mu_2 = a^2 \rho$
$\mu_3 = 2a^3 \rho$.

estimators: ml

$$\Sigma'(x - \hat{m}) = N\hat{a}\hat{\rho}$$

$$\Sigma' \ln(x - \hat{m}) = N(\psi(\hat{\rho}) + \ln \hat{a}), \quad \left(\psi(x) = \frac{d \ln \Gamma(x)}{dx}\right)$$

$$\Sigma'(x - \hat{m})^{-1} = N(\hat{a}/(\hat{\rho} - 1)).$$

moment

$$a^* = m_3/(2m_2) \quad \rho^* = 4m_2^3/m_3^2$$
$$m^* = m_1 - 2m_2^2/m_3.$$

Table 4.4: Bias and covariances

ml estimators: asymptotic variances and covariances, ρ large

$$\text{Var}_1 \hat{\rho} \sim 6(\rho - 1)^3 = \nu_{11}$$
$$\text{Var}_1 \hat{m} \sim 3a^2(\rho - 1)^3(\rho - 2)/2\rho = \nu_{22}$$
$$\text{Var}_1 \hat{a} \sim 3a^2(\rho - 1)/2 = \nu_{33}$$
$$\text{Cov}_1 (\hat{\rho}, \hat{m}) \sim -3a(\rho - 2)(\rho - 1)^2 = \nu_{12}$$
$$\text{Cov}_1 (\hat{\rho}, \hat{a}) \sim -3a(\rho - 1)^2 = \nu_{13}$$
$$\text{Cov}_1 (\hat{m}, \hat{a}) \sim 3a^2(\rho - 2)(\rho - 1)^2/2\rho = \nu_{23}.$$

(These are N^{-1} terms in the moments for $\rho \gg 0$).

COMMENTS

(a) Define $\hat{A} = (\hat{a} - a)/a$, $\hat{M} = (\hat{m} - m)/a$; then the estimating equations become (with $y = (x - m)/a$)

$$\Sigma(y - \hat{M}) = N\hat{\rho}(\hat{A} + 1)$$
$$\Sigma \ln(y - \hat{M}) = N\{\psi(\hat{\rho}) + \ln(\hat{A} + 1)\}$$
$$\Sigma(y - \hat{M})^{-1} = N/\{(\hat{\rho} - 1)(\hat{A} + 1)\}.$$

But the distribution of y is a function of ρ only. Hence the moments of $\hat{\rho}$, $(\hat{a} - a)/a$, $(\hat{m} - m)/a$ are functions of ρ, N only, which is of considerable use in reducing tabulations.

(b) Evaluations bring out the largeness of the N^{-2} coefficients in comparison with the N^{-1}; to dilute this effect would require unusually large sample sizes. For the one-fifth rule to be satisfied, critical sample sizes N^* for covariances are high, whereas the biases require higher values still; for example, at $\rho = 6$, a sample of 20000 or so would be needed to control the bias of $\hat{a}$.

(c) First-order biases and covariances are very unreliable, and higher-order terms will far outweigh the initial ones, even for samples as large as 1000; this is in marked contrast to the behaviour in the simultaneous estimation of $\hat{\rho}, \hat{a}$, m known, to be referred to in Chapter 6. The usual asymptotic efficiency, either for marginal distribution parameter estimation or for simultaneous

Table 4.4
Three-parameter Gamma distribution

BIAS

ρ	$E\,4(\hat{a}-a)/2a$ N^{-1}	N^{-2}	$E\,2(\hat{p}-\rho)/9\rho^2$ N^{-1}	N^{-2}	$E-2(\hat{m}-m)/3\rho^2$ N^{-1}	N^{-2}
5	2·17	—	−0·35	—	−0·83	—
6	1·78	−7836	−0·02	2855	−0·41	3722
7	1·58	−2278	0·17	790	−0·17	1025
8	1·47	−1090	0·30	381	0·00	480
9	1·39	−654	0·39	242	0·12	291
10	1·34	−447	0·46	186	0·22	206
15	1·20	−156	0·65	128	0·49	109
20	1·14	−94	0·74	147	0·62	108

COVARIANCES

ρ	$\dfrac{\text{Var }\hat{\rho}}{v_{11}}$ N^{-1}	N^{-2}	$\dfrac{\text{Var }\hat{m}}{v_{22}}$ N^{-1}	N^{-2}	$\dfrac{\text{Var }\hat{a}}{v_{33}}$ N^{-1}	N^{-2}	$\dfrac{\text{Cov}(\hat{p},\hat{m})}{v_{12}}$ N^{-1}	N^{-2}	$\dfrac{\text{Cov}(\hat{p},\hat{a})}{v_{13}}$ N^{-1}	N^{-2}	$\dfrac{\text{Cov}(\hat{m},\hat{a})}{v_{23}}$ N^{-1}	N^{-2}	N^*
5	1·013	434	1·079	501	1·095	392	1·012	480	1·012	401	1·161	564	2820
6	1·008	29	1·064	228	1·075	187	1·008	233	1·008	195	1·130	245	1225
7	1·005	41	1·053	165	1·061	127	1·006	181	1·005	143	1·108	162	905
8	1·004	63	1·046	145	1·051	99	1·004	169	1·004	125	1·093	129	845
9	1·003	97	1·040	140	1·045	83	1·003	171	1·003	119	1·082	113	855
10	1·002	222	1·036	142	1·039	73	1·002	179	1·002	119	1·073	105	1110
15	1·001	328	1·023	182	1·025	52	1·001	250	1·001	145	1·047	102	1640
20	1·001	447	1·017	236	1·018	44	1·001	334	1·001	183	1·035	117	2235

(N^* is the least sample size to make the second-order term one-fifth the first-order term *for all covariances for a given* ρ. Where there are no entries in the bias table, the machine output was considered to be unreliable.)

estimation, should be viewed with the greatest suspicion, unless accompanied by a caveat with respect to sample size.

(d) N^{-3} and higher-order terms in this case seem to be beyond the scope of current digital facilities.

REFERENCE

1. Bowman, K.O. and Shenton, L.R. (1969). "Remarks on maximum likelihood estimators for the gamma distribution". *Proc. Int. Conf. on Quality Control*, Tokyo, pp. 519–22.

NOTES ON THE REFERENCE

Results are given for the distribution of $\hat{\rho}$ when a is known ($m = 0$), including the first four moments of $\hat{\rho}$ to low orders in N^{-1}. For the 3-parameter case, ten choices of estimators are suggested. The N^{-1} covariance matrix is given for ml estimators, along with exact expressions for the N^{-1} biases. A short tabulation of the N^{-2} biases and covariances is given (reproduced here in Table 4.4) and these were found digitally using IBM 360-91 computer facility.

4.6 Two-parameter Hermite distribution (H.D.)

pgf: $\exp(a(t-1) + b(t^2-1))$, $a, b > 0$

probability function: $P_x = (\exp(-a-b)) \sum\sum_{i+2j=x} \dfrac{a^i b^j}{i!\, j!}$, $(x = 0, 1, \ldots)$ (4.7)

Examples:

$$P_0 = \exp(-a-b)$$

$$P_1 = aP_0$$

$$P_2 = \frac{a^2}{2!}\left(1 + \frac{2b}{a^2}\right) P_0$$

$$P_3 = \frac{a^3}{3!}\left\{1 + 3\left(\frac{2b}{a^2}\right)\right\} P_0$$

$$P_4 = \frac{a^4}{4!}\left\{1 + 6\left(\frac{2b}{a^2}\right) + 3\left(\frac{2b}{a^2}\right)^2\right\} P_0$$

. . . . . .

(The relation to the Hermite system of orthogonal polynomials $H_0(x) = 1$, $H_1(x) = x$, $H_2(x) = x^2 - 1$, $H_3(x) = x^3 - 3x$, $H_4(x) = x^4 - 6x^2 + 3$ is evident.)

moments: mean $= a + 2b$, variance $= a + 4b$

recurrence: $\begin{cases} P_{x+1} = (aP_x + 2bP_{x-1})/(x+1) & (x = 0, 1, \ldots) \\ P_x = 0, \quad x < 0. \end{cases}$

estimators: ml

$$\begin{cases} \sum_{x=0}^{\infty} n_x \dfrac{\hat{P}_{x-1}}{\hat{P}_x} = N \\ \sum_{x=0}^{\infty} n_x \dfrac{\hat{P}_{x-2}}{\hat{P}_x} = N, \end{cases}$$

where n_x is the frequency of the observation x,
$\hat{P}_x \equiv P_x$ when $a = \hat{a}$, $b = \hat{b}$.

(The second equation can be replaced by $\hat{a} + 2\hat{b} = m_1$, and then $\hat{a}$ and $\hat{b}$ can be found iteratively (for example, by using the method of scoring – cf. Rao, 1965 – ref. 43 in Chapter 1, page 302).

expectation identities:

$$E_s(\hat{a} + 2\hat{b}) = (a + 2b)\delta_{s,0}$$

$$\text{Var}_s(\hat{a}) + 4\,\text{Cov}_s(\hat{a}, \hat{b}) + 4\,\text{Var}_s(\hat{b}) = (a + 4b)\delta_{s,1}.$$

Table 4.5: Bias and covariances

COMMENTS

(a) In general, the second-order terms increase numerically as b increases, although the behaviour is not always quite straightforward. For example, $E_2\,\hat{a}$ decreases at first and ultimately increases as b increases. A glance at Table 4.5 shows that very often the second-order variances and covariances are significantly larger than the first-order terms. Roughly, N^{-2} terms play an important role, in order of decreasing importance, in E $\hat{a}$, Var $\hat{a}$, Cov $(\hat{a}, \hat{b})$, Var $\hat{b}$. Very large samples may be required whenever **b** exceeds **a** and **a** > 2. For example, if $a = 2$, $b = 5$, samples of 16000 or so are needed to damp out the second-order terms for the covariances, and 8000 or so for the biases. An impression of the N^{-2} terms can be gained from Fig. 4.23 (pages 116–17).

(b) Moment estimators, $a^* = 2m_1 - m_2$, $b^* = (m_2 - m_1)/2$, are quite simple to evaluate from data, and also are tractable from the sampling distribution viewpoint; clearly, infinite series in N^{-1} are not involved for all moments of finite orders. For example (Patel, 1971),

$$E\,a^* = a + A/N \qquad (a + 4b = A)$$

$$E\,b^* = b - A/2N$$

$$\mu_2(a^*) = (a + 2A^2)/N + (2a - A^2)/N^2 + (a + 16b)/N^3$$

Table 4.5
Bias and covariances for Hermite distribution

b	Bias ($\hat{a}$)	Var ($\hat{a}$)	Cov ($\hat{a}, \hat{b}$)	Var ($\hat{b}$)	Bias ($\hat{a}$)	Var ($\hat{a}$)	Cov ($\hat{a}, \hat{b}$)	Var ($\hat{b}$)
		$a = 0.1$				$a = 0.5$		
0.1	0.12 00	0.12 00	−0.01 00	0.11 00	1.0 00	1.3 00	−0.4 00	0.3 00
	0.30 00	0.40 00	−0.12 00	0.03 00	2.5 01	1.3 01	−5.5 00	2.4 00
0.5	0.12 00	0.12 00	−0.01 00	0.51 00	1.4 00	1.5 00	−0.5 00	0.8 00
	0.26 00	0.40 00	−0.12 00	0.03 00	1.8 01	2.2 01	−9.6 00	4.0 00
1.0	0.12 00	0.12 00	−0.01 00	1.01 00	1.5 00	1.6 00	−0.5 00	1.3 00
	0.26 00	0.40 00	−0.12 00	0.03 00	1.7 01	2.5 01	−1.1 01	4.6 00
5.0	−	−	−	−	1.6 00	1.6 00	−0.5 00	5.3 00
	−	−	−	−	1.7 01	2.8 01	−1.2 01	5.2 00
		$a = 1.0$				$a = 2.0$		
0.5	4.7 00	7.6 00	−3.3 00	2.2 00	5.5 00	2.7 01	−1.3 01	6.8 00
	8.9 01	1.9 02	−8.7 01	4.0 01	−9.6 01	2.1 01	−6.0 01	0.7 00
1.0	7.5 00	9.7 00	−4.4 00	3.2 00	1.2 01	5.0 01	−2.4 01	1.3 01
	3.3 02	5.1 02	−2.4 02	1.2 02	−6.5 02	7.2 01	−2.5 01	6.7 00
2.0	1.0 01	1.1 01	−5.2 00	4.6 00	3.6 01	1.1 02	−5.5 01	3.0 01
	6.2 02	9.8 02	−4.7 02	2.3 02	−3.7 03	3.9 03	−1.9 03	9.3 02
3.0	1.1 01	1.2 01	−5.5 00	5.8 00	7.6 01	1.8 02	−9.1 01	4.9 01
	7.5 02	1.2 03	−5.9 02	2.9 02	−2.8 03	3.0 04	−1.5 04	7.4 03
4.0	1.2 01	1.2 01	−5.7 00	6.8 00	1.3 02	2.5 02	−1.3 02	6.7 01
	8.3 02	1.4 03	−6.6 02	3.2 02	1.8 04	1.1 05	−5.4 04	2.7 04
5.0	1.2 01	1.3 01	−5.8 00	7.9 00	1.8 02	3.2 02	−1.6 02	8.4 01
	8.8 02	1.4 03	−7.0 02	3.4 02	7.3 04	2.5 05	−1.3 05	6.3 04
		$a = 3.0$				$a = 4.0$		
0.5	5.8 00	4.8 01	−2.3 01	1.2 01	6.5 00	7.2 01	−3.4 01	1.7 01
	8.5 02	−1.7 01	1.3 01	−8.4 00	3.2 01	−1.1 01	1.0 01	−7.3 00
1.0	9.4 00	8.5 01	−4.1 01	2.2 01	9.4 00	1.2 02	−5.8 01	3.0 01
	−6.0 01	−2.8 02	1.5 02	−7.6 01	2.5 01	−2.3 02	1.2 02	−6.3 01
2.0	1.9 01	2.0 02	−9.8 01	5.1 01	1.5 01	2.6 02	−1.3 02	6.5 01
	−8.8 02	−2.4 02	1.2 03	−6.0 02	1.9 01	−1.2 03	5.9 02	−3.0 02
3.0	3.1 01	3.7 02	−1.8 02	9.4 01	2.1 01	4.5 02	−2.2 02	1.2 02
	−4.7 03	−1.1 04	5.3 03	−2.6 03	−8.3 01	−3.1 03	1.6 03	−7.8 02
4.0	4.9 01	6.0 02	−3.0 02	1.5 02	2.7 01	7.1 02	−3.5 02	1.8 02
	−1.7 04	−3.5 04	1.8 04	−8.8 03	−4.6 02	−6.7 03	3.4 03	−1.7 03
5.0	7.4 01	8.8 02	−4.4 02	2.2 02	3.2 01	1.0 03	−5.2 02	2.6 02
	−4.6 04	−9.4 04	4.7 04	−2.4 04	−1.5 03	−1.3 04	6.7 03	−3.3 03

(The first and second entries correspond to the N^{-1} and N^{-2} terms respectively in the moment concerned.)

$$\text{Cov}(a^*, b^*) = -A^2/N + (-\tfrac{1}{2}a + 4b + A^2)/N^2 - (\tfrac{1}{2}a + 8b)/N^3$$

$$\mu_2(b^*) = (b + \tfrac{1}{2}A^2)/N + (-4b - \tfrac{1}{2}A^2)/N^2 + (a/4 + 4b)/N^3.$$

Exact higher moments can also be evaluated.

Some examples of the skewness and kurtosis of a^*, b^* are given in Table 4.6 on page 118.

Asymptotic variances of the moment estimators will of course exceed those of ml estimators. For example, there are the following comparisons.

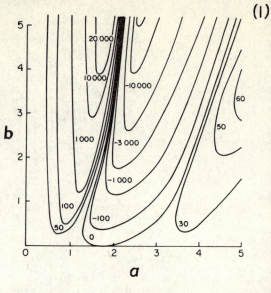

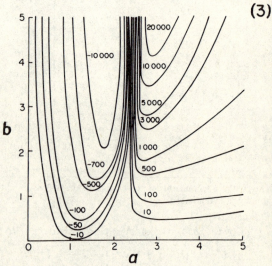

Fig. 4.23 Hermite distribution: N^{-2} terms in

$$a = 0.1 \qquad b = 0.1 \qquad N = 1$$

$$\mathrm{E}(\hat{a}-a) \sim 0.122 + 0.301 \qquad \mathrm{E}(\hat{b}-b) \sim -0.061 - 0.150$$

$$\mathrm{E}(a^*-a) = 0.500 \qquad \mathrm{E}(b^*-b) = -0.250$$

$$\mathrm{Var}\,\hat{a} \sim 0.123 + 0.395$$

$$\mathrm{Var}\,a^* = 0.600 - 0.300 + 1.700$$

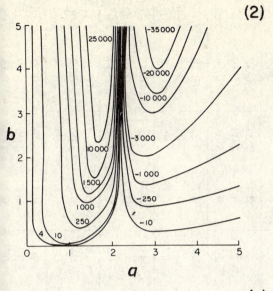

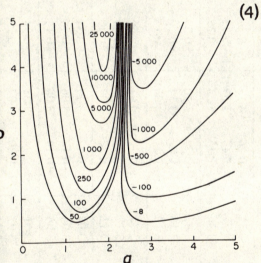

(1) bias $\hat{a}$, (2) Var $\hat{a}$, (3) Cov$(\hat{a}, \hat{b})$, (4) Var $\hat{b}$

$$\text{Cov}(\hat{a}, \hat{b}) \sim -0.011 - 0.123$$
$$\text{Cov}(a^*, b^*) = -0.250 + 0.600 - 0.850$$
$$\text{Var } \hat{b} \sim 0.106 + 0.025$$
$$\text{Var } b^* = 0.225 - 0.525 + 0.425$$

Table 4.6
Skewness and kurtosis of moment estimators for the Hermite distribution: sample sizes (i) 100, (ii) 300, (iii) 500

b	a		0.1		0.5		1.0		3.0		5.0	
			$\sqrt{\beta_1}$	β_2	$\sqrt{\beta_1}$	β_2	$\sqrt{\beta_1}$	β_2	$\sqrt{\beta_1}$	β_2	$\sqrt{\beta_1}$	β_2
0.1	a^*	(i)	−0.72	5.19	−0.20	3.44	−0.15	3.29	−0.18	3.15	−0.21	3.13
		(ii)	−0.43	3.75	−0.12	3.19	−0.09	3.10	−0.11	3.05	−0.12	3.05
		(iii)	−0.34	3.45	−0.10	3.11	−0.07	3.06	−0.08	3.03	−0.09	3.03
	b^*	(i)	1.16	5.32	0.76	4.16	0.54	3.66	0.36	3.27	0.32	3.21
		(ii)	0.67	3.77	0.44	3.39	0.32	3.22	0.21	3.09	0.19	3.07
		(iii)	0.53	3.46	0.34	3.23	0.25	3.13	0.16	3.05	0.14	3.04
1.0	a^*	(i)	−0.41	3.47	−0.37	3.40	−0.34	3.34	−0.29	3.22	−0.28	3.18
		(ii)	−0.24	3.16	−0.22	3.14	−0.20	3.11	−0.17	3.08	−0.16	3.06
		(iii)	−0.18	3.10	−0.17	3.08	−0.16	3.07	−0.13	3.05	−0.12	3.04
	b^*	(i)	0.52	3.53	0.49	3.47	0.46	3.41	0.38	3.28	0.35	3.23
		(ii)	0.30	3.18	0.28	3.16	0.26	3.14	0.22	3.09	0.20	3.08
		(iii)	0.24	3.11	0.22	3.09	0.20	3.08	0.17	3.06	0.15	3.05
5.0	a^*	(i)	−0.31	3.09	−0.31	3.19	−0.31	3.19	0.33	3.19	0.32	3.18
		(ii)	−0.18	3.06	−0.18	3.06	−0.18	3.06	0.19	3.06	0.19	3.06
		(iii)	−0.14	3.04	−0.14	3.04	−0.14	3.04	0.15	3.04	0.14	3.04
	b^*	(i)	0.34	3.21	0.34	3.21	0.34	3.20	0.33	3.19	0.32	3.18
		(ii)	0.20	3.07	0.20	3.07	0.19	3.07	0.19	3.06	0.19	3.06
		(iii)	0.15	3.04	0.15	3.04	0.15	3.04	0.15	3.04	0.14	3.04

$$a = 0.5 \quad b = 0.1 \quad N = 1$$

$E(\hat{a}-a) \sim 0.980 + 24.602$ $\qquad$ $E(\hat{b}-b) \sim -0.490 - 12.301$

$E(a^*-a) = 0.900$ $\qquad$ $E(b^*-b) = -0.450$

$\operatorname{Var} \hat{a} \sim 1.333 + 12.722$

$\operatorname{Var} a^* = 2.120 - 0.620 + 2.100$

$\operatorname{Cov}(\hat{a}, \hat{b}) \sim -0.417 - 5.530$

$\operatorname{Cov}(a^*, b^*) = -0.810 + 0.960 - 1.050$

$\operatorname{Var} \hat{b} \sim 0.308 + 2.350$

$\operatorname{Var} b^* = 0.505 - 0.805 + 0.525.$

$$a = 1.5 \quad b = 2.0 \quad N = 1$$

$E(\hat{a}-a) \sim 31 + 3117$ $\qquad$ $E(\hat{b}-b) \sim -15 - 1559$

$E(a^*-a) \sim 9.50$ $\qquad$ $E(b^*-b) \sim -4.75$

$\operatorname{Var} \hat{a} \sim 49 + 7668$

$\operatorname{Var} a^* \sim 182 - 178 + 34$

$\operatorname{Cov}(\hat{a}, \hat{b}) \sim -24 - 3769$

$\operatorname{Cov}(a^*, b^*) \sim -90 + 98 - 17$

$\operatorname{Var} \hat{b} \sim 14 + 1853$

$\operatorname{Var} b^* \sim 47 - 53 + 8$

$$a = 2 \quad b = 2 \quad N = 1$$

$E(\hat{a}-a) \sim 36 - 3736$ $\qquad$ $E(\hat{b}-b) \sim -18 + 1868$

$E(a^*-a) \sim 10$ $\qquad$ $E(b^*-b) \sim -5$

$\operatorname{Var} \hat{a} \sim 113 + 3915$

$\operatorname{Var} a^* \sim 202 - 196 + 34$

$\operatorname{Cov}(\hat{a}, \hat{b}) \sim -55 - 1910$

$\operatorname{Cov}(a^*, b^*) \sim -100 + 107 - 17$

$\operatorname{Var} \hat{b} \sim 30 + 931$

$\operatorname{Var} b^* \sim 52 - 58 + 9.$

Notice that second-order terms for the covariances of mles are not in general less numerically than corresponding terms for moment estimators.

(c) *Recommendations for users* Patel (1971) has given a comprehensive study of various estimators for the Hermite distribution, including moment, ml and linear functions of frequencies. We confine attention to ml and moment estimators. The latter are simple to compute, simple to simulate, and have uncomplicated moments, for they are linear functions of m_1, m_2.

As for ml, we suggest the following points should be noted, after having decided on the approximate values of a, b either from historical evidence or from the data.

(i) *Biases* The N^{-2} term in E $\hat{a}$ can outweigh the N^{-1} term in a region approximated by $a \geqslant 0.6$ and inside a parabolic arc passing through $(0.6, 0.3)$, $(3, 2)$, $(4, 5)$; in this region $N > 100$ at least is required.

For the bias of b, use the relation $\hat{a} + 2\hat{b} = m'_1$, so that, for example, $E_1 \hat{a} = -2E_1 \hat{b}$.

(ii) *Covariances* The "bad" region for Var $\hat{a}$ and Cov$(\hat{a}, \hat{b})$ is inside the region bounded by $a \geqslant 0.7$ and a parabolic arc passing through $(0.7, 0.4)$, $(3, 3)$, and $(4, 5)$. Samples exceeding a hundred may be needed here. For Var $\hat{b}$, there is a similar (narrower) region of large-sample requirement, bounded by $a \geqslant 1$ and a parabolic arc passing through $(1, 0.5)$, $(2.5, 3)$ and $(3.2, 5)$.

(iii) Rough assessments of sample size may be deduced from Table 4.7. In general, beware whenever $b > a$.

REFERENCES

1. Kemp, A.W. and Kemp, C.D. (1965). "Some properties of the 'Hermite' distribution." *Biometrika*, 52, 381–94.
2. Patel, Y.C. (1971). "Some problems in estimation for the parameters of the Hermite distribution." Thesis in partial fulfilment of Ph.D. degree requirements, Univ. of Georgia, Athens, Georgia.
3. Patel, Y.C., Shenton, L.R. and Bowman, K.O. (1974). "Maximum likelihood estimation for the parameters of the Hermite distribution." *Sankhyā*, 36, B, Pt 2, 154–62.
4. Rao, C.R. (1952). *Advanced Statistical Methods in Biometric Research.* John Wiley & Sons, Inc., New York.

NOTES ON THE REFERENCES

(1) defines the Hermite distribution in its most general form and for the two-parameter case shows the relation to Hermite orthogonal polynomials. Models leading to the distribution are given and ml estimation is discussed with examples.

(2) discusses in detail several estimators for the two-parameter Hermite distribution and gives their low-order moments. These are exact for moment estimators, and include terms of the second order for the bias and covariances of ml estimators. Four moment approximations to the distributions of moment estimators are included.

Table 4.7

Safe sample sizes for controlling the second-order term in the bias of $\hat{a}$ and covariances for the Hermite distribution (N^{-2} coefficient reduced to 5 percent of N^{-1} coefficient)

b	a						
	0·1	0·5	1·0	2·0	3·0	4·0	5·0
0·1	50	503	62	—	—	—	—
	65	191	141	—	—	—	—
	224	266	169	—	—	—	—
	5	153	139	—	—	—	—
0·5	43	254	383	350	77	99	116
	65	294	485	16	8	4	1
	226	377	521	10	12	6	3
	1	107	373	3	15	9	5
1·0	42	230	889	1121	111	53	70
	65	325	1053	29	65	38	26
	226	414	1118	21	71	42	29
	1	74	728	11	71	43	30
2·0	—	—	1225	2103	178	25	61
	—	—	1726	695	238	91	53
	—	—	1820	690	243	94	55
	—	—	998	628	236	93	55
5·0	—	—	1455	7888	4414	915	24
	—	—	2311	16003	2147	257	66
	—	—	2427	16047	2153	258	67
	—	—	861	15035	2104	254	66

(Gaps in this table arise because of the demands on computing time required for the evaluation of the second-order terms. The entries in the table refer, in order, to bias of $\hat{a}$, Var $\hat{a}$, Cov($\hat{a}, \hat{b}$), and Var $\hat{b}$. For the bias of $\hat{b}$, note that $\hat{a} + 2\hat{b} = m'_1$.)

4.7 An exponential regression model

4.7.1 Cox and Snell (reference 3, Chapter 2) considered the estimation of the parameters a, b for the model

$$y_j = a\, e^{bd_j} \epsilon_j, \quad (j = 1, 2, \ldots, n), \tag{4.8}$$

where $d_j = x_j - \bar{x}$ and (ϵ_j) are independently and identically distributed exponential variates with unit mean. For the ml estimators $\hat{a}, \hat{b}$, using the likelihood function

$$\prod_{j=1}^{n} a^{-1} \exp(-bd_j) \cdot \exp(a^{-1} y_j \exp(-bd_j)),$$

we derive the equations

$$\Sigma y_j d_j \exp(-\hat{b}d_j) = 0$$

$$\Sigma y_j \exp(-\hat{b}d_j) = n\hat{a}.$$

Cox and Snell, using a neat approach (mentioned in Chapter 2), derive the first-order biases and covariances of $\hat{a}, \hat{b}$. It is, however, very tedious to go farther; we indicate briefly how higher-order terms can be found.

4.7.2 The equation for $\hat{b}$ may be written, using a summatory notation,

$$(y_j k_j^{(1)}) - (\hat{b} - b)(y_j k_j^{(2)}) + \frac{(\hat{b} - b)^2}{2!}(y_j k_j^{(3)}) - \ldots = 0,$$

where

$$k_j^{(s)} = d_j^s e^{-bd_j}.$$

Note that, defining $y_j = a(\epsilon_j + e^{bd_j})$ with $E \epsilon_j = 0$,

$$(y_j k_j^{(1)}) = \Sigma a(\epsilon_j + e^{bd_j})d_j e^{-bd_j}$$

$$= a(\epsilon_j d_j \delta_j), \quad (\delta_j = e^{-bd_j}),$$

so this term is of zero order in expectation, and the order of magnitude of $E(y_j k_j^{(1)})^s$ is n^{-r} where $r = [(s+1)/2]$.

But from Lagrange's expansion (section 2.5) we have, for the first power, for example,

$$\hat{b} - b = c_1 + c_1^2 c_3/2 + c_1^3(3c_3^2 - c_4)/6 +$$
$$+ c_1^4(c_5 - 10c_3c_4 + 15c_3^3)/24 + \ldots,$$

where

$$c_s = \frac{D_s + \theta_s}{D_2 + \theta_2} \quad (D_s = \Sigma d_j^s, \quad \theta_s = (\epsilon d^s \delta))$$

and in particular $c_1 = \theta_1/(D_2 + \theta_2)$.

Expectations of products of θ_s terms must now be evaluated. For example,

$$E \theta_1 = 0, \quad \text{Var } \theta_1 = D_2, \quad \mu_3(\theta_1) = 2D_3,$$

$$\mu_4(\theta_1) = 6D_4 + 3D_2^2, \quad \mu_5(\theta_1) = 24D_5 + 20D_2D_3,$$

$$\mu_6(\theta_1) = 120D_6 + 90D_4D_2 + 40D_3^2 + 15D_2^3;$$

$$E \theta_1\theta_2 = D_3, \quad E \theta_1\theta_2^2 = 2D_5, \quad E \theta_1\theta_2^3 = 6D_7 + 3D_3D_4,$$

$$E \theta_1\theta_2^4 = 24D_9 + 12D_4D_5 + 8D_3D_6;$$

$$E \theta_1^3\theta_3 = 6D_6 + 3D_2D_4, \quad E \theta_1^3\theta_2 = 6D_5 + 3D_2D_3,$$

$$E \theta_1^2\theta_2\theta_3 = 6D_7 + 2D_3D_4 + D_2D_5,$$

etc.

The D's play a similar role to sample size in assessing relative importance.

Thus, for order of magnitude,

$$\frac{D_3^l D_4^m \cdots}{D_2^r} = O\{n^{-(r-l-m-\cdots)}\},$$

so that, for example, $D_3/D_2^2 = O(1/n)$, $D_3 D_4/D_2^4 = O(1/n^2)$, $D_3^2 D_5/D_2^6 = O(1/n^3)$.

4.7.3 Terms in the Lagrange expansion of a power of $\hat{b}$ (or $\hat{b} - b$), in expectation, are forced into a hierarchy of magnitudes by the first term θ_1. Thus,

$$E\, c_1^3 c_4 = \left\langle \frac{E\, \theta_1^3 (D_4 + \theta_4)}{(D_2 + \theta_2)^4} \right\rangle$$

$$= \left\langle \frac{E(\theta_1^3 D_4 + \theta_1^3 \theta_4)}{D_2^4} \left(1 - \frac{4\theta_2}{D_2} + \frac{10\theta_2^2}{D_2^2} - \frac{20\theta_2^3}{D_2^3} + \cdots \right) \right\rangle$$

from which terms of various orders of magnitude can be extracted. After considerable algebraic manipulation, we find the following moments and moment ratios (terms displayed in order of magnitude):

$$E\, \hat{b} \sim b + \left\{-\frac{1}{2}\frac{D_3}{D_2^2}\right\} + \left\{\frac{9}{8}\frac{D_5}{D_2^3} - \frac{31 D_3 D_4}{12 D_2^4} + \frac{11}{8}\frac{D_3^3}{D_2^5}\right\} \quad (4.9)$$

$$+ \left\{-\frac{215}{48}\frac{D_7}{D_2^4} + \left(\frac{205}{16}\frac{D_3 D_6}{D_2^5} + \frac{565}{48}\frac{D_4 D_5}{D_2^5}\right)\right.$$

$$\left. - \left(\frac{275}{12}\frac{D_3 D_4^2}{D_2^6} + \frac{95}{4}\frac{D_3^2 D_5}{D_2^6}\right) + \left(\frac{75}{2}\frac{D_3^3 D_4}{D_2^7} - \frac{175}{16}\frac{D_3^5}{D_2^8}\right)\right\},$$

$$\mathrm{Var}\, \hat{b} \sim \frac{1}{D_2} + \left(\frac{D_4}{D_2^3} - \frac{D_3^2}{2 D_2^4}\right) + \left(-\frac{9 D_6}{4 D_2^4} + \frac{35}{12}\frac{D_4^2}{D_2^5} + \frac{3 D_3 D_5}{D_2^5}\right.$$

$$\left. -\frac{17}{4}\frac{D_3^2 D_4}{D_2^6} + \frac{3 D_3^4}{4 D_2^7}\right),$$

$$\mu_3(\hat{b}) \sim \frac{-D_3}{D_2^3} + \left(\frac{9}{2}\frac{D_5}{D_2^4} - \frac{12 D_3 D_4}{D_2^5} + \frac{13 D_3^3}{2 D_2^6}\right),$$

$$\mu_4(\hat{b}) \sim \frac{3}{D_2^2} + \left(\frac{8 D_4}{D_2^4} - \frac{3 D_3^2}{D_2^5}\right) + \left(-\frac{39}{2}\frac{D_6}{D_2^5} + \frac{12 D_3 D_5}{D_2^6} + \frac{65 D_4^2}{2 D_2^6}\right.$$

$$\left. -\frac{21}{2}\frac{D_3^2 D_4}{D_2^7} - \frac{39}{4}\frac{D_3^4}{D_2^8}\right);$$

$$\begin{cases} \sqrt{\beta_1}(\hat{b}) \sim -D_3/D_2^{3/2}, \\ \beta_2(\hat{b}) \sim 3 + \left(\frac{2 D_4}{D_2^2}\right) + \left(-\frac{6 D_6}{D_2^3} - \frac{6 D_3 D_5}{D_2^4} + \frac{8 D_4^2}{D_2^4} + \frac{20 D_3^2 D_4}{D_2^5} - \frac{15 D_3^4}{D_2^6}\right). \end{cases}$$

Similarly for $\hat{a}$, we have:

$$E\frac{\hat{a}}{a} \sim 1 + \left(-\frac{1}{2n}\right) + \left(\frac{D_4}{8nD_2^2} - \frac{5D_3^2}{24nD_2^3}\right) + \left(\frac{3D_3D_5}{nD_2^4} - \frac{4D_6}{nD_2^3}\right),$$

$$\text{Var}\left(\frac{\hat{a}}{a}\right) \sim \frac{1}{n} + \left(-\frac{1}{2n^2}\right) + \left\{\frac{3D_3D_5}{n^2D_2^4} - \frac{4D_6}{n^2D_2^3} - \frac{1}{n^2}\left(\frac{D_4}{8D_2^2} - \frac{5}{24}\frac{D_3^2}{D_2^3}\right)\right\}^2,$$

$$\mu_3\left(\frac{\hat{a}}{a}\right) \sim \frac{2}{n^2} + \left(-\frac{1}{n^3} + \frac{21D_4}{8n^2D_2^2} - \frac{27}{8n^2}\frac{D_3^2}{D_2^3}\right),$$

$$\mu_4\left(\frac{\hat{a}}{a}\right) \sim \frac{3}{n^2} + \frac{3}{n^3} + \left(\frac{5D_3^2}{4n^3D_2^3} - \frac{3D_4}{4n^3D_2^2} - \frac{9}{4n^4}\right);$$

$$\begin{cases} \sqrt{\beta_1(\hat{a})} \sim \frac{2}{\sqrt{n}}\left\{1 + \left(\frac{1}{4n} + \frac{21D_4}{16D_2^2} - \frac{27D_3^2}{16D_2^3}\right)\right\} \\ \beta_2(\hat{a}) \sim 3 + \frac{6}{n} + \left(\frac{3}{n^2} + \frac{5D_3^2}{4nD_2^3} - \frac{3D_4}{4nD_2^2}\right) \end{cases}$$

For the leukaemia data cited by Cox and Snell, we find for the moments (showing contributions in order of magnitude):

$$D_1 = 0 \quad D_2 = 6\cdot2608 \quad D_3 = -0\cdot0972 \quad D_4 = 5\cdot0041$$
$$D_5 = -1\cdot1210 \quad D_6 = 5\cdot2183.$$

Moments of $\hat{b}$

$$E(\hat{b} - b) \sim 0\cdot1597 + 0\cdot0204 - 0\cdot0000$$
$$= 0\cdot1801$$
$$\text{Var } \hat{b} \sim 0\cdot1597 + 0\cdot0004 - 0\cdot0027$$
$$= 0\cdot1574$$
$$\sqrt{\beta_1(\hat{b})} \sim -0\cdot0062$$
$$\beta_2(\hat{b}) \sim 3 + 0\cdot2553 + 0\cdot0025$$
$$= 3\cdot2578.$$

Moments of $\hat{a}$

$$E\frac{\hat{a}}{a} \sim 1 - 0\cdot0294 + 0\cdot0009 - 0\cdot0050$$
$$= 0\cdot9665$$
$$\text{Var }\frac{\hat{a}}{a} \sim 0\cdot0588 - 0\cdot0017 - 0\cdot0007$$
$$= 0\cdot0564$$

$$\sqrt{\beta_1}(\hat{a}) \sim 0.4851 + 0.0071 + 0.0812$$
$$= 0.5734$$
$$\beta_2(\hat{a}) \sim 3 + 0.3529 + 0.0048$$
$$= 3.3577.$$

Evidently, higher-order terms of the moments of $\hat{a}$ and $\hat{b}$ are almost negligible. It is of interest to note that if we introduce the skewness and kurtosis measures for the x's, namely $\sqrt{b_1} = m_3/m_2^{3/2}$, $b_2 = m_4/m_2^2$, then $\sqrt{\beta_1}(\hat{b}) \sim -\sqrt{b_1}$, $\beta_2(\hat{b}) \sim 3 + 2b_2/n$, and $\sqrt{\beta_1}(\hat{a}) \sim (2/\sqrt{n})(1 + (4 + 21b_2 - 27b_1)/16n)$, $\beta_2(\hat{a}) \sim 3 + 6/n + (12 + 5b_1 - 3b_2)/(4n^2)$. Clearly higher-order terms could be quite large.

5 SPECIAL DENSITY ESTIMATION[*]

5.1 Introduction

In Chapter 1, *1.3.4*(b), we referred to estimators given implicitly in terms of a fixed number of statistics, this number being independent of the sample size. The general formulae of Chapters 2 and 3 can still be used as checks on the N^{-1} and N^{-2} coefficients in biases and covariances, but now terms as high as N^{-6} (or higher) for the first four to six moments may be attainable by special techniques and devices. Clearly no general guidelines can be set out.

The examples chosen are such that the Taylor series involved are low-dimensioned, and this is the main reason for the gain in simplicity.

5.2 Logarithmic series distribution (L.S.D.)

5.2.1 This distribution has many applications in ecology as a model for the distribution of species abundance, and in meteorology as a model for the duration (in days or other units) of rainfall storms (Shenton and Skees, 1969). In 1964, Patil, Kamat and Wani (also Patil and Wani, 1965) used expressions described in Chapter 2 to evaluate the N^{-1} and N^{-2} coefficients in the bias and variance of the mle. This was extended to the third and fourth moments, including terms of order N^{-9}, by the present authors (1970).

5.2.2 Basic formulae

If X is a logarithmic series random variate, then its probability function can be written

$$\Pr(X = x) = \alpha \theta^x / x,$$

where $x = 1, 2, \ldots$; $\alpha^{-1} = \ln 1/(1 - \theta)$, $0 \leqslant \theta < 1$, and the maximum likelihood estimator of θ is a solution of the equation

$$\frac{\hat{\theta}}{(1 - \hat{\theta}) \ln (1/(1 - \hat{\theta}))} = \bar{x}, \tag{5.1}$$

where $\bar{x}$ is the mean of the random sample $x_1, x_2, \ldots, x_N$. The solution $\hat{\theta}$ is unique, and in particular when $\bar{x} = 1$, $\hat{\theta} = 0$.

The stochastic Taylor expansion for $\hat{\theta}$ in terms of the random incremental variables $x_1 = \bar{x} - E(\bar{x})$ is

---

[*] Estimation of the parameters of the logarithmic series, zero-truncated binomial, and two-parameter gamma distributions are considered.

$$\hat{\theta} = \theta + \sum_{r=1}^{\infty} \frac{x_1^r \bar{\theta}_r}{r!}, \tag{5.2}$$

where

$$\bar{\theta}_r = \left.\frac{d^r \hat{\theta}}{d\bar{x}^r}\right|_{\hat{\theta}=\theta,\, \bar{x}=\mu}$$

$$\mu = E\bar{x} = \frac{-\theta}{(1-\theta)\ln(1-\theta)}.$$

We require the central moments (ν_s) of the mean $\bar{x}$, and we put

$$\nu_s = E(\bar{x} - E\bar{x})^s$$

$$\mu_s = E(x - \mu)^s, \quad \text{where } \mu = Ex \text{ and } \mu_0 = 1.$$

The values of ν_s can be set up recursively from the relation

$$\sum_{r=0}^{s} \binom{s}{r} \nu_{s+1-r} \mu_r / N^r = \mu_{s+1}/N^s, \quad s = 1, 2, \ldots, \nu_1 = 0. \tag{5.3}$$

5.2.3 Central moments, μ_s

We have, if

$$\mu_s' = E x^s, \tag{5.4}$$

then

$$E e^{\lambda x} = -\alpha \ln(1 - \theta e^{\lambda}), \quad |\theta e^{\lambda}| < 1. \tag{5.5}$$

Define

$$-\ln(1 - \theta e^{\lambda}) = \sum_{s=0}^{\infty} \lambda^s T_s(\theta)/s! \tag{5.6}$$

$$= T(\lambda, \theta), \text{ say,}$$

where, for example, $T_0(\theta) = \alpha^{-1}$, $T_1(\theta) = \theta/(1-\theta)$.

Clearly,
$$\frac{\partial T(\lambda, \theta)}{\partial \lambda} = \frac{\theta \partial T(\lambda, \theta)}{\partial \theta} \tag{5.7}$$

so that
$$T_{s+1}(\theta) = \theta \frac{dT_s(\theta)}{d\theta}, \quad s = 0, 1, \ldots. \tag{5.8}$$

Also from (5.6), differentiating with respect to θ and λ, it follows that

$$T_s(\theta) = \theta + \theta \sum_{r=0}^{s-1} \binom{s-1}{r} T_{r+1}(\theta), \quad s = 1, 2, \ldots. \tag{5.9}$$

Formula (5.8) can be used for algebraic expressions for $T_s(\theta)$ using Formac language; (5.9) can be used to set up, at least theoretically, any number of $T(\theta)$'s. Then $\mu_s' = \alpha T_s(\theta)$.

A few examples are:

$$T_1(\theta) = \frac{\theta}{1-\theta}$$

$$T_2(\theta) = \frac{\theta}{(1-\theta)^2}$$

$$T_3(\theta) = \frac{\theta + \theta^2}{(1-\theta)^3} \qquad (5.10)$$

$$T_4(\theta) = \frac{\theta + 4\theta^2 + \theta^3}{(1-\theta)^4}$$

$$T_5(\theta) = \frac{\theta + 11\theta^2 + 11\theta^3 + \theta^4}{(1-\theta)^5}$$

$$T_6(\theta) = \frac{\theta + 26\theta^2 + 66\theta^3 + 26\theta^4 + \theta^5}{(1-\theta)^6}.$$

The numerators in the $T(\theta)$'s are well known in analysis, the coefficients being Eulerian numbers, the polynomials those defined by Riordan (1958) as $A_{m,1}(\theta)$ (see his Table 4, p. 215 and Example 2, p. 38).

For the central moments of X, we have

$$\mu_s = \frac{1}{T_0} \sum_{r=0}^{s} (-1)^r \binom{s}{r} \mu^r T_{s-r}(\theta)$$

leading to the formula,

$$\mu_s = \frac{1}{\ln(1-\theta)} \left(\theta \frac{d}{d\theta} - \mu\right)^s \ln(1-\theta),$$

where μ is replaced by $\alpha T_1(\theta)$ after differentiation. Numerically, for programming purposes, equations similar to (5.10) are preferred. Define

$$T_s(\theta) = (1-\theta)^{-s} \sum_{-\infty}^{\infty} a_r^{(s)} \theta^r, \quad s = 1, 2, \ldots, \qquad (5.11)$$

where $a_r^{(s)} = 0, \quad r \leq 0, r \geq s;$

$$= 1, \quad r = 1.$$

Then from (5.9) it follows that

$$a_r^{(s+1)} = r a_r^{(s)} + (s-r+1) a_{r-1}^{(s)}, \quad s = 1, 2, \ldots. \qquad (5.12)$$

This bivariate recursive relation has been used by us to set up the central moments of X as far as μ_{18}, sufficient to evaluate expectations of moments as far as terms to order N^{-9}. A similar relation to (5.12) was given by Patil et al. (1966).

5.2.4 Derivatives of $\hat{\theta}$

The expansion for $\hat{\theta}$ could be set up from the Lagrange expansion given in (2.8). For example,

$$\Gamma = \ln \alpha + x \ln \theta - \ln x, \quad (\alpha \ln(1-\theta) = -1)$$

and $\Gamma_s = \dfrac{d^s \Gamma}{d\theta^s}$

$$= \frac{(-1)^{s-1} x}{\theta^s} + A_s,$$

where $A_s = \dfrac{d^s \alpha}{d\theta^s}$.

Thus

$$A_1 = \frac{-\alpha}{1-\theta},$$

$$A_2 = \frac{\alpha^2 - \alpha}{(1-\theta)^2},$$

$$A_3 = \frac{-2\alpha^3 + 3\alpha^2 - 2\alpha}{(1-\theta)^3},$$

$$A_4 = \frac{6\alpha^4 - 12\alpha^3 + 11\alpha^2 - 6\alpha}{(1-\theta)^4},$$

and in general, if $A_s = \pi_s(\alpha)/(1-\theta)^s$, then

$$A_{s+1} = (-\alpha^2 \pi_s^{(1)}(\alpha) + s\pi_s(\alpha))/(1-\theta)^{s+1}.$$

Again $(n\Gamma_s) = A_s + \dfrac{(-1)^{s-1}}{\theta^s} \bar{x},$

and the elements in (2.8) become

$$d_s = \frac{A_s + \dfrac{(-1)^{s-1} \mu}{\theta^s} + \dfrac{(-1)^{s-1} \epsilon}{\theta^s}}{A_2 - \dfrac{\mu}{\theta^2} - \dfrac{\epsilon}{\theta^2}}, \quad (\epsilon = \bar{x} - \mu).$$

After inverting the denominators, an expansion for $\hat{\theta}$ in terms of $\bar{x} - \mu$ will be found, not without, however, some tiresome algebra.

The Taylor series (5.2) is to be seen then as a rearrangement of the Lagrange expansion. If a general formula can be found for the derivatives, the moment series problem will have been solved.

We require $d^r \hat{\theta}/d\bar{x}^r$, where

$$f(\hat{\theta}) = 1/\bar{x}, \tag{5.13}$$

and $f(\theta) = -(1/\theta - 1)\ln(1 - \theta)$. Evidently we must evaluate $d^m f(\theta)/d\theta^m$ and use Faà di Bruno's formula to deal with $d^r\theta/d\bar{x}^r$. Now

$$\theta f^{(1)}(\theta) + f(\theta) = \frac{-\theta f(\theta)}{1-\theta} + 1$$

or

$$\theta(1-\theta)f^{(1)}(\theta) + f(\theta) = 1 - \theta, \tag{5.14a}$$

from which

$$\theta(1-\theta)f^{(m+1)}(\theta) + \{1 + m(1-\theta)\}f^{(m)}(\theta) - m(m-1)f^{(m-1)}(\theta)$$
$$= -\delta_{m,1} + (1-\theta)\delta_{m,0}, \quad m = 0, 1, \ldots, \tag{5.14b}$$

where $f^{(0)}(\theta) = f(\theta)$, $\delta_{r,s}$ is the Kronecker delta function. The programming of (5.14), (5.12), (5.11) and (5.3), using $(\hat{\theta} - \theta)^s = (\hat{\theta} - \theta)^{s-1} * (\hat{\theta} - \theta)$, leads to series in descending powers of N for the moments (the expression (5.3) has to be modified to filter out powers of N^{-1}).

5.2.5 Comments on the moments

Coefficients of powers of N^{-1} up to N^{-9} for the first four moments of $\hat{\theta}$ are given in Tables 5.1A to 5.1C and are graphically displayed up to N^{-8} in Fig. 5.1 to 5.4. Convergence is most rapid for μ_1 and becomes progressively retarded for μ_2, μ_3, and μ_4. Moreover, as θ increases towards unity, the series for each moment becomes less stable; this is brought out by noticing the series for $\theta = 0.99, N = 9, N = 10$, and particularly for μ_4. In fact, it is clear that if we take N large enough to damp off the terms in the series for μ_4, then N will also be large enough to damp off the terms in the lower moments. As an overall remark,[*] convergence (using the word in a broad sense) appears to be satisfactory for $\theta \leq 0.9$ and $N \geq 9$ (the case $0.9 < \theta < 1.0$ would need larger values of N for stabilization, particularly as $\theta \to 1$). This can be seen from the terms in the fourth moment for $\theta = 0.9$, which are:

Terms in fourth moment: $\theta = 0.9$

N	8	9	10	12
N^{-2}	·000054	·000048	·000035	·000024
N^{-3}	·000287	·000227	·000147	·000085
N^{-4}	·000623	·000438	·000255	·000122
N^{-5}	·000705	·000440	·000231	·000093
N^{-6}	·000276	·000153	·000072	·000024
N^{-7}	−·000385	−·000190	−·000081	−·000023
N^{-8}	−·000588	−·000258	−·000099	−·000023
N^{-9}	−·000017	−·000007	−·000002	·000000

[*] Further comments on summing the moment series are given in Appendix B.

Table 5.1A

Logarithmic series distribution: Terms in expansions in powers of N^{-1} of moments of $\hat{\theta}$ ($N = 8$)

	$\theta = 0.1$	$N = 8$			$\theta = 0.5$	$N = 8$		
	μ_1	μ_2	μ_3	μ_4	μ_1	μ_2	μ_3	μ_4
N^{-1}	−19482	20967	–	–	−66203	38867	–	–
N^{-2}	3293	−6830	2762	1319	3477	5409	−7271	4532
N^{-3}	−460	1140	−297	−1252	1142	−4413	1902	2793
N^{-4}	39	46	−618	1162	−359	−40	2869	−3766
N^{-5}	5	−115	447	−626	2	718	−1345	−1065
N^{-6}	−4	48	−151	85	34	−192	−708	2811
N^{-7}	1	−11	2	171	−11	−104	795	−308
N^{-8}	0	−1	34	−183	−1	93	2	−1837
N^{-9}	0	2	−25	94	3	−7	−398	1095

	$\theta = 0.8$	$N = 8$			$\theta = 0.95$	$N = 8$		
	μ_1	μ_2	μ_3	μ_4	μ_1	μ_2	μ_3	μ_4
N^{-1}	−55863	12800	–	–	−23400	1302	–	–
N^{-2}	−5682	9612	−2655	492	−6913	1981	−120	51
N^{-3}	1505	1721	−3155	1692	−700	1547	−336	356
N^{-4}	313	−1273	−636	1951	384	623	−469	109
N^{-5}	−160	−513	1222	218	126	−54	−374	2040
N^{-6}	−22	300	646	−1437	−45	−217	−96	2440
N^{-7}	29	158	−529	−816	−27	−69	155	1488
N^{-8}	−1	−117	−442	−1011	9	70	185	−692
N^{-9}	−7	−57	322	1034	8	58	6	−2485

(Entries in Tables 5.1A to 5.1C must be multiplied by 10^{-6})

Table 5.1B
L.S.D.: Terms in expansions in powers of N^{-1} of moments of $\hat{\theta}$ ($N = 9$)

	$\theta = 0.01$ $N = 9$				$\theta = 0.1$ $N = 9$			
	μ_1	μ_2	μ_3	μ_4	μ_1	μ_2	μ_3	μ_4
N^{-1}	−1840	2185	—	—	−17317	18638	—	—
N^{-2}	310	−792	461	14	2602	−5396	2183	1042
N^{-3}	−49	200	−241	78	−323	801	−209	−879
N^{-4}	7	−42	76	−50	24	29	−386	726
N^{-5}	−1	7	−17	14	3	−64	248	−347
N^{-6}	0	−1	2	0	−2	24	−75	42
N^{-7}	0	0	0	−2	0	−5	0	75
N^{-8}	0	0	0	2	0	0	13	−71
N^{-9}	0	0	0	0	0	0	−9	33

	$\theta = 0.4$ $N = 9$				$\theta = 0.7$ $N = 9$			
	μ_1	μ_2	μ_3	μ_4	μ_1	μ_2	μ_3	μ_4
N^{-1}	−53182	37673	—	—	−57978	20134	—	—
N^{-2}	4111	71	−4466	4258	−1904	8845	−4090	1216
N^{-3}	344	−2861	3001	12	1324	−543	−2269	2435
N^{-4}	−198	572	592	−2456	−18	−1115	947	852
N^{-5}	28	180	−999	1028	−86	108	898	−1202
N^{-6}	6	−135	198	729	16	198	−307	−745
N^{-7}	−4	18	213	−845	7	−49	−335	674
N^{-8}	0	19	−159	67	−4	−44	162	493
N^{-9}	0	−13	3	413	0	24	136	−495

	$\theta = 0.95$ $N = 9$				$\theta = 0.99$ $N = 9$			
	μ_1	μ_2	μ_3	μ_4	μ_1	μ_2	μ_3	μ_4
N^{-1}	−20800	1158	—	—	−5935	65	—	—
N^{-2}	−5462	1565	−95	4	−2612	133	−2	0.01
N^{-3}	−491	1087	−236	25	−770	160	−6	0.12
N^{-4}	240	389	−293	68	−68	134	−13	0.52
N^{-5}	70	−30	−208	113	59	78	−20	1.53
N^{-6}	−22	−107	−47	120	23	21	−23	3.38
N^{-7}	−12	−30	68	65	−6	−11	−19	5.93
N^{-8}	4	27	72	−27	−5	−16	−8	8.24
N^{-9}	3	20	2	−86	0	−4	4	8.56

Table 5.1C
L.S.D.: Terms in expansions in powers of N^{-1} of moments of $\hat{\theta}$ ($N = 10$)

	$\theta = 0.01$ $N = 10$				$\theta = 0.1$ $N = 10$			
	μ_1	μ_2	μ_3	μ_4	μ_1	μ_2	μ_3	μ_4
N^{-1}	−1656	1967	—	—	−15586	16774	—	—
N^{-2}	251	−642	374	12	2107	−4371	1768	844
N^{-3}	−36	146	−175	57	−236	584	−152	−641
N^{-4}	5	−27	50	−33	16	19	−253	476
N^{-5}	−1	4	−10	9	2	−37	146	−21
N^{-6}	0	−1	1	0	0	13	−40	2
N^{-7}	0	0	0	0	0	−2	0	4
N^{-8}	0	0	0	0	0	0	1	−3
N^{-9}	0	0	0	0	0	0	0	1

	$\theta = 0.4$ $N = 10$				$\theta = 0.7$ $N = 10$			
	μ_1	μ_2	μ_3	μ_4	μ_1	μ_2	μ_3	μ_4
N^{-1}	−47864	33905	—	—	−52181	1812	—	—
N^{-2}	3330	57	−3617	3449	−1542	716	−3313	985
N^{-3}	251	−2086	2188	9	968	−40	−1654	1775
N^{-4}	−130	376	388	−1611	−12	−73	621	559
N^{-5}	17	106	−590	607	−51	6	530	−709
N^{-6}	3	−71	105	387	8	11	−163	−396
N^{-7}	−2	9	102	−404	3	−2	−160	322
N^{-8}	0	8	−68	29	−2	−2	70	212
N^{-9}	0	−5	1	160	0	1	53	−192

	$\theta = 0.95$ $N = 10$				$\theta = 0.99$ $N = 10$			
	μ_1	μ_2	μ_3	μ_4	μ_1	μ_2	μ_3	μ_4
N^{-1}	−18720	1042	—	—	−5341	58	—	—
N^{-2}	−4424	1268	−77	3	−2116	11	−1	0.01
N^{-3}	−358	792	−172	18	−561	116	−4	0.08
N^{-4}	157	255	−192	45	−45	88	−9	0.34
N^{-5}	41	−18	−123	67	35	46	−12	0.90
N^{-6}	−12	−57	−25	64	12	11	−12	1.80
N^{-7}	−6	−14	33	31	−3	−5	−9	2.83
N^{-8}	2	12	31	−12	−2	−7	−4	3.54
N^{-9}	1	8	8	−33	0	−2	2	3.32

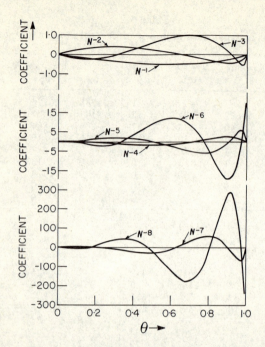

Fig. 5.1 Logarithmic series distribution: N^{-1} through N^{-8} coefficients in bias of $\hat{\theta}$

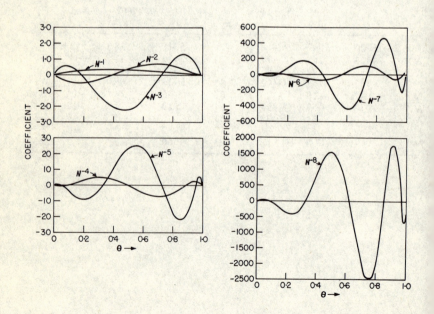

Fig. 5.2 L.S.D.: N^{-1} through N^{-8} coefficients in Var $\hat{\theta}$

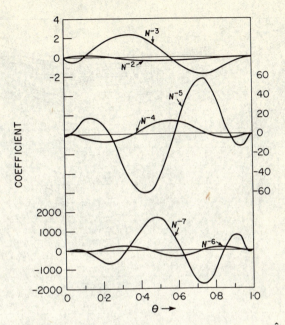

Fig. 5.3 L.S.D.: N^{-2} through N^{-7} coefficients in $\mu_3(\hat{\theta})$

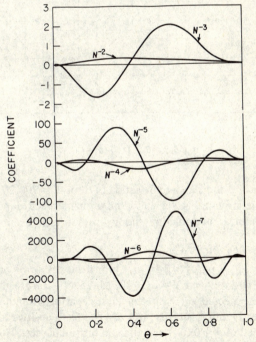

Fig. 5.4 L.S.D.: N^{-2} through N^{-7} coefficients in $\mu_4(\hat{\theta})$

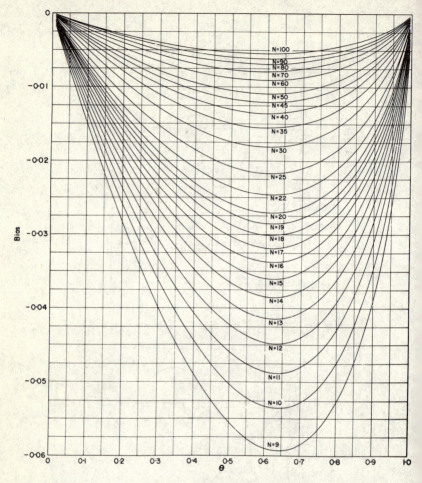

Fig. 5.5 L.S.D.: $E(\hat{\theta} - \theta)$

Graphs of the bias and standard deviation of $\hat{\theta}$ for $N = 9(1)20, 22, 25$ $(5)50(10)100$ are given in Fig. 5.5 and 5.6. For many practical purposes, these, along with linear interpolation, should give sufficient accuracy.

5.2.6 Skewness and kurtosis

The values of $\sqrt{\beta_1} = \mu_3/\mu_2^{3/2}$, $\beta_2 = \mu_4/\mu_2^2$ for the distribution of $\hat{\theta}$ for values of $\theta = 0.1(0.1)0.9$ and $N = 8, 9, 10, 15, 25, 50, 100$ are given in Table 5.2. As might be expected, there is the greatest departure from normality for N small and θ nearly unity; actually, $\sqrt{\beta_1}$ changes from positive to negative in the neighbourhood of $\theta = 0.35$, depending on the sample size. However, very marked departures from normality would occur for $\theta > 0.9$ and for most

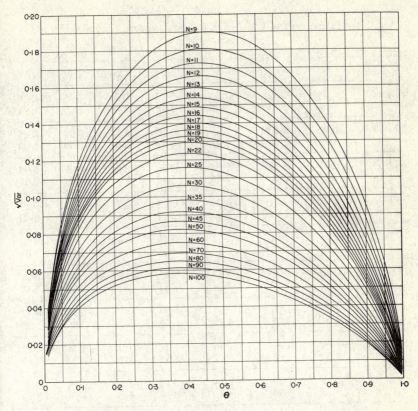

Fig. 5.6 L.S.D.: Standard deviation of $\hat{\theta}$

sample sizes, and this would correspond to a situation in which small samples were drawn from a long-tailed distribution.

A graphical display of the skewness and kurtosis is given in Fig. 5.7 to 5.9.

5.2.7 Verification of the results

The moments and moment parameters $\sqrt{\beta_1}, \beta_2$ can be evaluated by a sample configuration method, provided the distribution sampled has a short tail (θ small). This has been computerized for $\theta = 0.1$, truncating the series at $x = 9$ and carrying computation to 16 digits. All sample configurations are set up; for each, the value of $\hat{\theta}$ is derived by an iterative procedure using (5.1) in the form

$$f^*(\hat{\theta}) = \frac{-(1-\hat{\theta})}{\hat{\theta}} \ln(1-\hat{\theta}) - \frac{1}{\bar{x}}$$

with a starting value $\hat{\theta} = \dfrac{9\bar{x} - 6 - \sqrt{(9\bar{x}^2 - 12\bar{x} + 6)}}{6\bar{x} - 2}$ (derived from a rational

137

Table 5.2
L.S.D.: Skewness and Kurtosis: $\sqrt{\beta_1}$ and β_2 for maximum likelihood estimator $\hat{\theta}$

θ	N	8	9	10	15	25	50	100
0.1	$\sqrt{\beta_1}$	1.14	1.06	1.00	0.77	0.57	0.38	0.26
	β_2	3.32	3.15	3.05	2.88	2.85	2.89	2.94
0.2	$\sqrt{\beta_1}$	0.48	0.43	0.39	0.26	0.16	0.09	0.05
	β_2	1.93	2.14	2.23	2.43	2.62	2.80	2.90
0.3	$\sqrt{\beta_1}$	0.10	0.06	0.03	−0.05	−0.09	−0.09	−0.07
	β_2	2.17	2.20	2.25	2.48	2.69	2.85	2.98
0.4	$\sqrt{\beta_1}$	−0.22	−0.24	−0.26	−0.29	−0.28	−0.23	−0.17
	β_2	2.73	2.54	2.52	2.69	2.87	2.96	2.99
0.5	$\sqrt{\beta_1}$	−0.51	−0.51	−0.52	−0.51	−0.45	−0.35	−0.26
	β_2	2.62	2.71	2.80	3.03	3.12	3.11	3.07
0.6	$\sqrt{\beta_1}$	−0.76	−0.77	−0.77	−0.73	−0.62	−0.47	−0.34
	β_2	2.37(*)	2.97	3.24	3.52	3.47	3.30	3.17
0.7	$\sqrt{\beta_1}$	−1.04	−1.06	−1.06	−0.97	−0.81	−0.59+	−0.42+
	β_2	4.04(*)	4.25	4.33	4.29	3.98	3.56	3.29
0.8	$\sqrt{\beta_1}$	−1.54++	−1.50+	−1.46+	−1.28+	−1.03+	−0.74+	−0.52+
	β_2	8.02(*)	6.96	6.47	5.61	4.76	3.92	3.47
0.9	$\sqrt{\beta_1}$	−2.38(*)	−2.24	−2.14	−1.78+	−1.36+	−0.94+	−0.65+
	β_2	7.50(*)	8.94(*)	9.43(†)	8.48	6.34	4.58	3.76

(*) Not reliable because of slow rate of series convergence.
(†) Slightly suspect because of slow convergence.
All distributions in this table are Pearson Type I, excepting +Pearson Type VI and ++Pearson Type IV.

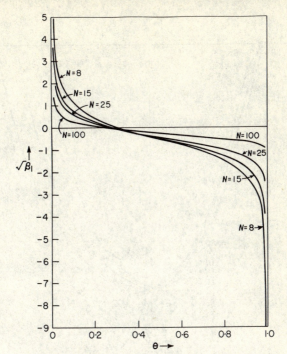

Fig. 5.7 L.S.D.: Skewness, $\sqrt{\beta_1}(\hat{\theta})$

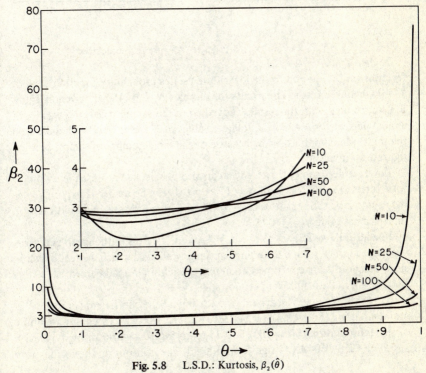

Fig. 5.8 L.S.D.: Kurtosis, $\beta_2(\hat{\theta})$

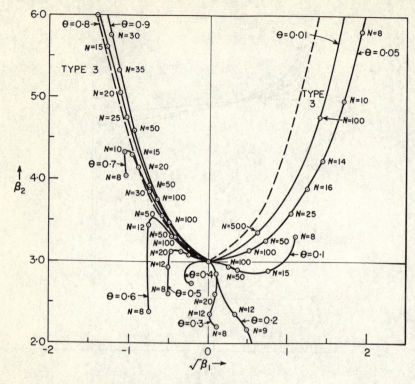

Fig. 5.9 L.S.D.: $\sqrt{\beta_1}(\hat{\theta}), \beta_2(\hat{\theta})$ for various sample sizes

fraction approximation to the logarithmic series probability function), taking $\hat{\theta} = 0$ whenever $\bar{x} = 1$. The results for samples $N = 2(1)10$ are given in Table 5.3. Agreement with the results derived from the asymptotic series was almost perfect for $N = 8, 9, 10$; larger values of N are difficult to handle from the multinomial point of view because of the large number of sample configurations involved.

Lower-order moments for $N < 8$ can also be checked against the asymptotic series (Table 5.1C, with $N = 10$, $\theta = 0.1$, can be used). Thus, for $N = 5$, the asymptotic series leads to $\mu_1 = 0.075631$, in good agreement with the multinomial value $\mu_1 = 0.07564$.

In addition, we remark that our N^{-1} and N^{-2} terms in the mean and variance agree (apart from a difference in sign in the bias, which we feel is a slip, since our results are confirmed otherwise) to within a small percentage with those given in Patil and Wani (1965, p. 402).

As for sheer numerical accuracy in setting up the coefficients in the moments using (5.2), (5.3), (5.5), (5.13) and (5.14), we carried out a section of the computations on an IBM 360-75 (carrying 16 significant digits) and also on a CDC 1604A (24 significant digits). The printout on the former gave

Table 5.3

L.S.D.: Moments of $\hat{\theta}$ using (a) multinomial methods, (b) asymptotic series[*]

$$\theta = 0.1$$

Multinomial probabilities

$p_1 = \cdot 94912216$	$p_4 = \cdot 00023728$	$p_7 = \cdot 00000014$
$p_2 = \cdot 04745611$	$p_5 = \cdot 00001898$	$p_8 = \cdot 00000001$
$p_3 = \cdot 00316374$	$p_6 = \cdot 00000158$	$p_9 = \cdot 00000000$

N	μ_1	μ_2	μ_3	μ_4	$\sqrt{\beta_1}$	β_2
2	0.05464	0.02743	0.01248	0.00659	2.748	8.757
3	0.06472	0.02532	0.00864	0.00381	2.145	5.949
4	0.07117	0.02273	0.00613	0.00242	1.790	4.690
5	0.07564	0.02038	0.00452	0.00167	1.552	4.023
6	0.07892	0.01837	0.00344	0.00123	1.381	3.631
7	0.08142	0.01668	0.00269	0.00094	1.251	3.386
8	0.08339	0.01525	0.00216	0.00075	1.148	3.225
	(0.08339)	(0.01525)	(0.00215)	(0.00077)	(1.144)	(3.316)
9	0.08499	0.01403	0.00177	0.00061	1.065	3.116
	(0.08499)	(0.01403)	(0.00177)	(0.00062)	(1.063)	(3.149)
10	0.086303	0.012978	0.001473	0.000512	0.996	3.040
	(0.086303)	(0.012978)	(0.001472)	(0.000514)	(0.996)	(3.053)

[*] The entries in this table are those derived by a "multinomial" approach; those in parentheses are derived from the asymptotic series.

Table 5.4

L.S.D.: Comparison of accuracy for two digital machines for moments of $\hat{\theta}$

		$\theta = 0.1$			
Moment	Machine	Coefficient			
		N^{-6}	N^{-7}	N^{-8}	N^{-9}
μ_1'	CDC	$-9\cdot 80296 - 01$	$2\cdot 46162\ \ 00$	$-3\cdot 75021\ \ 00$	$-6\cdot 57008 - 01$
	IBM	$-9\cdot 80302 - 01$	$2\cdot 46081\ \ 00$	$-3\cdot 87420\ \ 00$	$-2\cdot 20258 - 01$
μ_2	CDC	$1\cdot 27122\ \ 01$	$-2\cdot 27021\ \ 01$	$-1\cdot 35073\ \ 01$	$3\cdot 01380\ \ 02$
	IBM	$1\cdot 27122\ \ 01$	$-2\cdot 27019\ \ 01$	$-1\cdot 34827\ \ 01$	$3\cdot 05616\ \ 02$
μ_3	CDC	$-3\cdot 97107\ \ 01$	$4\cdot 21796\ \ 00$	$5\cdot 72816\ \ 02$	$-3\cdot 39896\ \ 02$
	IBM	$-3\cdot 97107\ \ 01$	$4\cdot 21794\ \ 00$	$5\cdot 72813\ \ 02$	$-3\cdot 39953\ \ 03$
μ_4	CDC	$2\cdot 21946\ \ 01$	$3\cdot 59080\ \ 02$	$-3\cdot 06399\ \ 03$	$1\cdot 26392\ \ 04$
	IBM	$2\cdot 21943\ \ 01$	$3\cdot 59080\ \ 02$	$-3\cdot 06399\ \ 03$	$1\cdot 26393\ \ 04$
		$\theta = 0.2$			
μ_1'	CDC	$-2\cdot 52675\ \ 00$	$1\cdot 71147\ \ 00$	$1\cdot 07325\ \ 01$	$-5\cdot 70469\ \ 01$
	IBM	$-2\cdot 52675\ \ 00$	$1\cdot 71147\ \ 00$	$1\cdot 07325\ \ 01$	$-5\cdot 70520\ \ 01$
μ_2	CDC	$5\cdot 63836\ \ 00$	$6\cdot 90363\ \ 01$	$-3\cdot 64310\ \ 02$	$7\cdot 81936\ \ 02$
	IBM	$5\cdot 63836\ \ 00$	$6\cdot 90363\ \ 01$	$-3\cdot 64310\ \ 02$	$7\cdot 81938\ \ 02$

six digits, and on the latter seven digits. For $\theta \geqslant 0.1$ the agreement in all terms was nearly perfect. A selection of comparisons is given in Table 5.4.

Discrepancies do occur for small values of θ ($\theta < 0.1$) and particularly for the lower-order moments. For example, the coefficients of the N^{-1} through N^{-9} terms in the first two moments for the IBM and CDC machines were as follows:

$$\theta = 0.05$$

Coefficient	Mean value of $\hat{\theta}$		Variance of $\hat{\theta}$	
	IBM	CDC	IBM	CDC
N^{-1}	−8·06632−02	−8·06632−02	9·17997−02	9·18000−02
N^{-2}	1·16520−01	1·16520−01	−2·73259−01	−2·73259−01
N^{-3}	−1·51019−01	−1·51019−01	5·17114−01	5·17114−01
N^{-4}	1·64656−01	1·64656−01	−6·08932−01	−6·08932−01
N^{-5}	−1·08726−01	−1·08734−01	−1·88604−01	−1·88603−01
N^{-6}	−1·24954−01	−1·26906−01	3·78485 00	3·78504 00
N^{-7}	7·44028−01	2·08863−01	−1·38225 01	−1·37693 01
N^{-8}	−2·04014 00	−1·69692 02	3·26555 01	4·93348 01
N^{-9}	4·00048 00	−5·90814 04	−3·83587 01	5·84326 03

It will be seen that for the mean, agreement is good up to the N^{-4} term and is almost completely lost from the N^{-7} term onwards. The agreement for the variance is a little better, but is completely lost at the N^{-9} term. Agreement for the third and fourth moments improves, and for the latter there is agreement in the first digit for the highest coefficient. A possible explanation for the improved accuracy for the higher moments relates to the fact that the first moment involves an eighteenth derivative (and moment), the second moment involves a seventeenth derivative (and moment), and so on. These high-order terms only affect the high-order terms in N^{-1}; obviously, because of the recursive schemes involved in the moments (doubly interlaced) and derivatives (Faà di Bruno's formula plus a second-order recurrence), higher-order terms involve many more sums of products than the lower-order terms. It is possible then that the inaccuracies appear through large summations of terms involving products of numbers of widely different orders of magnitude.

5.2.8 Conclusion and discussion

We have derived moment parameters for the distribution of $\hat{\theta}$, the maximum likelihood estimator of θ in the logarithmic series distribution. The mean, variance, skewness ($\sqrt{\beta_1}$) and kurtosis (β_2) are given for $0.1 \leqslant \theta \leqslant 0.9$ and $N \geqslant 8$; smaller values of θ could be dealt with using the sample configuration method, provided N is not too large. For small θ and large N the asymptotic series approach could be used, although "rounding off" errors and similar computer difficulties would need careful attention.

The interval $0.9 < \theta < 1.0$ presents serious difficulties for the asymptotic series approach, although moments could be assessed for large to very large values of N, the sample size.

As a general statement the distribution of $\hat{\theta}$ is not too far removed from the normal distribution; however, departures from normality become serious when θ exceeds about 0·9, when the sample size is less than about nine, and when both of these conditions are satisfied.

Verification of the results has been sought through (a) the use of two digital computers, (b) the use of a sample configuration method for small values of θ and small sample sizes.

REFERENCES

1. Bowman, K. O. and Shenton, L. R. (1970). "Properties of the maximum likelihood estimators for the parameter of the logarithmic series distribution". In *Random Counts in Models and Structures*, ed. G. P. Patil, Penn State Statistics Series.
2. Patil, G. P., Kamat, A. R., and Wani, J. K. (1966). "Certain studies on the structure and statistics of the logarithmic series distribution and related tables". *Tech. Rep. Aerospace Res. Lab.*, Wright-Patterson Air Force Base, p. 389.
3. Patil, G. P. and Wani, J. K. (1965). "Maximum likelihood estimation for the complete and truncated logarithmic series distribution". In *Classical and Contagious Discrete Distributions Symposium*, ed. G. P. Patil, Statistical Publishing Company, Calcutta (Pergamon Press, Oxford).
4. Riordan, J. (1958). *An Introduction to Combinatorial Analysis*. John Wiley & Sons, New York.
5. Shenton, L. R. and Skees, P. (1969). "Some statistical aspects of amounts and duration of rainfall". *Sampling and Other Aspects of Statistical Ecology*. Pennsylvania State Univ. Press, University Park.

5.3 Zero-truncated binomial distribution (Zt.B.D.)

5.3.1 Basic formulae

The ml estimator of p for the probability function

$$\Pr(X = x) = \binom{n}{x} p^x q^{n-x}/(1-q^n), \quad x = 1, 2, \ldots, n \qquad (5.15)$$

(n a positive integer; $p + q = 1$)

is the solution of

$$f(\hat{p}) = \frac{1}{\bar{x}},$$

where $f(p) = \dfrac{(1-q^n)}{n(1-q)}$ (Bowman and Shenton, 1971).

5.3.2 Taylor expansion

For $\hat{p}$ we use

$$\hat{p} = p + \sum_{s=1}^{\infty} \frac{(\bar{x} - \mu)^s}{s!} \frac{\partial^s p}{\partial \bar{x}^s}\bigg|_{\hat{p}=p,\,\bar{x}=\mu} \qquad \left(\mu = \frac{np}{1-q^n}\right) \tag{5.16}$$

in which the derivatives are computed using Faà de Bruno's formula. For this we need the expression

$$\frac{d^s f(\hat{p})}{d\hat{p}^s} = \frac{(-1)^s}{n} \sum_{r=0}^{n-1} r(r-1)\ldots(r-s+1)\hat{q}^{r-s}, \qquad (n = 2, 3, \ldots).$$

The central moments $\mu_s^* = E(\bar{x} - \mu)^s$ are deduced from the central moments of X, using the formula

$$\mu_{s+1} = -p \sum_{r=1}^{s} \binom{s}{r} \mu_{s-r+1} + \frac{npq}{1-q^n}((1-\mu)^s q^{n-1} - \mu_s)$$

$$+ npq \frac{(1-q^{n-1})}{1-q^n} \sum_{r=0}^{s} \binom{s}{r} \mu_r. \tag{5.17}$$

5.3.3 Tabulations

To illustrate the behaviour of the moments of the parent population, the derivatives of $\hat{\theta}$ with respect to $\hat{p}$ at $\hat{p} = p$, $\bar{x} = \mu_1'$ and properties of the distribution of $\hat{\theta}$, we give the following:

Table 5.5 Moments of the population and derivatives of $\hat{p}$
Table 5.6 Sampling moments of the mean
Table 5.7 N^{-1} through N^{-9} coefficients in the first four moments of $\hat{p}$
Table 5.8 Bias of $\hat{p}$
Table 5.9 Var $\hat{p}$
Table 5.10 Skewness and kurtosis of $\hat{p}$

5.3.4 Comments on the tables

(i) The moments μ_s increase moderately rapidly (Table 5.5).
(ii) The derivatives $(1/s!)\,\partial^s p/\partial \bar{x}^s$ at the origin ($\hat{p} = p$, $\bar{x} = \mu$) alternate in sign and decrease moderately rapidly in value (Table 5.5).
(iii) The sampling moments of $\bar{x}$ increase at about the same rate as the derivatives in (ii) decrease (Table 5.6).
(iv) The coefficients in the moment expansions of $\hat{p}$ are remarkably stable (Table 5.7). For example, the magnitude of the ratio of the N^{-9} term to the N^{-1} term for a given value of p rarely exceeds 10^4, so that sample sizes as small as 5 can be used.
(v) Tables 5.8 to 5.10 have been constructed with sample size N sufficiently large to damp out the importance of higher-order terms; the smallest N used is 10, which is a conservative value for damping purposes.

Table 5.5
Zero-truncated binomial distribution: Moments and derivatives
$n = 10 \qquad p = 0.25$

Central moments of random variable X (μ_1 = Mean)		Derivatives of $\hat{p}$ $s!\ p_s = \partial^s \hat{p}/\partial \bar{x}^s$ at $\hat{p} = p, \bar{x} = \mu_1'$	
s	μ_s	s	p_s
1	2·049 00	1	1·178–01
2	1·592 00	2	−9·633–03
3	1·210 00	3	3·334–03
4	7·762 00	4	−9·554–04
5	1·718 01	5	2·833–04
6	7·371 01	6	−9·060–05
7	2·626 02	7	2·967–05
8	1·128 03	8	−9·860–06
9	4·909 03	9	3·334–06
10	2·295 04	10	−1·141–06
11	1·117 05	11	3·941–07
12	5·681 05	12	−1·372–07
13	2·995 06	13	4·809–08
14	1·633 07	14	−1·695–08
15	9·169 07	15	6·003–09
16	5·289 08	16	−2·135–09
17	3·127 09	17	7·620–10
18	1·889 10	18	−2·729–10
19	1·165 11	19	9·800–11

As a general comment, the distribution of the maximum likelihood estimator is nearly normal, the greatest discrepancy occurring when $N = 10$, $n = 2, p = 0.1$ for which $\sqrt{\beta_1} = 1.0$, $\beta_2 = 3.14$.

(vi) Numerical checks are available for the bias given in Table 5.8 for selected values of N, n, p, for Thomas and Gart (1971) give a short table of the exact bias. Our results are in excellent agreement, with an error less than 2·0 percent. Further checks could be set up by using the relation

$$n\hat{p} = \bar{x}\{1 - (1 - \hat{p})^n\}$$

and its powers.

The greatest error in Tables 5.8 to 5.10 is thought to be approximately 2 percent, and accuracy improves for $0.2 < p < 0.8$, and n large. For $N \geq 50$ the error is perhaps less than 1 percent.

Table 5.6

Zt.B.D.: Sampling moments of the mean

$n = 10 \qquad p = 0.25$

(including only coefficients up to N^{-9})

s	$\mu_s(\bar{x})$		
2	$1 \cdot 591670^0(1)$		
3	$1 \cdot 210170^0(2)$		
4	$7 \cdot 600243^0(2)$	$6 \cdot 143233^{-2}(3)$	
5	$1 \cdot 926192^1(3)$	$-2 \cdot 086398^0(4)$	
6	$6 \cdot 048540^1(3)$	$1 \cdot 611182^1(4)$	$-2 \cdot 891476^0(5)$
7	$3 \cdot 219156^2(4)$	$-6 \cdot 713599^1(5)$	$7 \cdot 796453^0(6)$
8	$6 \cdot 739097^2(4)$	$6 \cdot 853687^2(5)$	$-2 \cdot 701259^2(6)$	$3 \cdot 867369^1(7)$
9	$6 \cdot 148601^3(5)$	$-1 \cdot 352656^3(6)$	$1 \cdot 366571^2(7)$	$-2 \cdot 344338^1(8)$
10	$9 \cdot 653778^3(5)$	$2 \cdot 415467^4(6)$	$-1 \cdot 454393^4(7)$	$4 \cdot 413397^3(8)$	$-7 \cdot 244262^2(9)$
11	$1 \cdot 345650^5(6)$	$-8 \cdot 334416^3(7)$	$-2 \cdot 168120^4(8)$	$8 \cdot 615207^3(9)$
12	$1 \cdot 690220^5(6)$	$8 \cdot 389868^5(7)$	$-6 \cdot 401680^5(8)$	$2 \cdot 747073^5(9)$
13	$3 \cdot 341256^6(7)$	$1 \cdot 157183^6(8)$	$-2 \cdot 603331^6(9)$	
14	$3 \cdot 497354^6(7)$	$3 \cdot 023172^7(8)$	$-2 \cdot 528866^7(9)$	
15	$9 \cdot 306127^7(8)$	$1 \cdot 010271^8(9)$		
16	$8 \cdot 349951^7(8)$	$1 \cdot 151075^9(9)$		
17	$2 \cdot 878027^9(9)$	—		
18	$2 \cdot 259363^9(9)$	—		

(The parenthetic entry is the coefficient of N^{-1} with which the entry is associated; the superscript is the power of 10 by which the entry must be multiplied.)

Table 5.7

Zt.B.D.: N^{-1} through N^{-9} coefficients in first four moments of $\hat{p}$

$n = 10 \qquad p = 0.25$

Moment	Coefficient								
	N^{-1}	N^{-2}	N^{-3}	N^{-4}	N^{-5}	N^{-6}	N^{-7}	N^{-8}	N^{-9}
$E(\hat{p} - p)$	$-1.533-02$	$-3.227-03$	$-8.153-05$	$8.534-04$	$9.955-04$	$6.635-04$	$-3.264-04$	$-2.180-03$	$-4.367-03$
$\mu_2(\hat{p})$	$2.209-02$	$3.693-03$	$2.051-04$	$-7.621-04$	$-1.185-03$	$-1.038-03$	$9.122-05$	$2.650-03$	$6.451-03$
$\mu_3(\hat{p})$	—	$-5.368-05$	$-1.652-04$	$3.734-04$	$8.552-04$	$9.390-04$	$2.488-04$	$-1.942-03$	$-6.118-03$
$\mu_4(\hat{p})$	—	$1.464-03$	$5.499-04$	$-8.487-05$	$-4.893-04$	$-7.590-04$	$-5.402-04$	$9.760-04$	$4.698-03$

Table 5.8
Zt.B.D.: Bias of $\hat{p}$ ($E(\hat{p} - p)$)

n	p	Sample size					
		10	20	30	40	50	100
2	0·1	−7·96−03	−4·13−03	−2·78−03	−2·10−03	−1·69−03	−8·49−04
	0·5	−1·89−02	−9·43−03	−6·27−03	−4·70−03	−3·76−03	−1·88−03
	0·9	−5·20−03	−2·54−03	−1·68−03	−1·25−03	−1·00−03	−4·97−04
6	0·1	−2·94−03	−1·49−03	−1·00−03	−7·52−04	−6·03−04	−3·02−04
	0·5	−1·58−03	−7·75−04	−5·14−04	−3·34−04	−3·07−04	−1·53−04
	0·9	−3·20−06	−1·26−06	−7·74−07	−5·55−07	−4·33−07	−2·05−07
10	0·1	−1·99−03	−1·00−03	−6·69−04	−5·02−04	−4·02−04	−2·01−04
	0·5	−1·91−04	−9·14−05	−6·00−05	−4·47−05	−3·56−05	−1·76−05
	0·9	−3·56−10	−3·50−10	−1·88−10	−1·26−10	−9·40−11	−4·09−11

Table 5.9
Zt.B.D.: Var $\hat{p}$

n	p	Sample size					
		10	20	30	40	50	100
2	0·1	1·42−02	7·59−03	5·18−03	3·93−03	3·16−03	1·60−03
	0·5	2·93−02	1·44−02	9·52−03	7·12−03	5·68−03	2·83−03
	0·9	6·07−03	2·87−03	1·88−03	1·40−03	1·11−03	5·50−04
6	0·1	2·80−03	1·42−03	9·52−04	7·16−04	5·73−04	2·87−04
	0·5	4·62−03	2·29−03	1·52−03	1·14−03	9·10−04	4·54−04
	0·9	1·50−03	7·50−04	5·00−04	3·75−04	3·00−04	1·50−04
10	0·1	1·44−03	7·21−04	4·81−04	3·61−04	2·89−04	1·45−04
	0·5	2·53−03	1·26−03	8·42−04	6·31−04	5·05−04	2·52−04
	0·9	9·00−04	4·50−04	3·00−04	2·25−04	1·80−04	9·00−05

Table 5.10
Zt.B.D.: Skewness and kurtosis of $\hat{p}$

n	p		Sample size					
			10	20	30	40	50	100
2	0·1	$\sqrt{\beta_1}$	1·03	0·66	0·53	0·45	0·40	0·28
		β_2	3·14	2·96	2·95	2·95	2·96	2·97
	0·5	$\sqrt{\beta_1}$	−0·37	−0·29	−0·24	−0·22	−0·19	−0·14
		β_2	2·93	3·02	3·03	3·02	3·02	3·01
	0·9	$\sqrt{\beta_1}$	−0·95	−0·66	−0·54	−0·47	−0·42	−0·29
		β_2	4·34	3·64	3·42	3·32	3·25	3·12
6	0·1	$\sqrt{\beta_1}$	0·29	0·19	0·15	0·13	0·12	0·08
		β_2	2·84	2·91	2·94	2·96	2·96	2·98
	0·5	$\sqrt{\beta_1}$	−0·10	−0·07	−0·05	−0·05	−0·04	−0·03
		β_2	3·03	3·02	3·01	3·01	3·01	3·00
	0·9	$\sqrt{\beta_1}$	−0·35	−0·24	−0·20	−0·17	−0·15	−0·11
		β_2	3·09	3·04	3·03	3·02	3·02	3·01
10	0·1	$\sqrt{\beta_1}$	0·15	0·10	0·08	0·07	0·06	0·04
		β_2	2·90	2·95	2·97	2·97	2·98	2·99
	0·5	$\sqrt{\beta_1}$	−0·02	−0·01	−0·01	−0·01	−0·01	−0·00
		β_2	3·00	3·00	3·00	3·00	3·00	3·00
	0·9	$\sqrt{\beta_1}$	−0·27	−0·19	−0·15	−0·13	−0·12	−0·08
		β_2	3·05	3·03	3·02	3·01	3·01	3·01

REFERENCES

1. Bowman, K. O. and Shenton, L. R. (1971). "Tables of moments of the maximum likelihood estimator of the parameter p for the zero-truncated binomial distribution". ORNL Report 4747.
2. Shenton, L. R., Bowman, K. O., and Sheehan, D. (1971). "Sampling moments of moments associated with univariate distributions". *J. Roy. Statist. Soc.*, (B), **33**, No. 3, 444–57.
3. Thomas, D. G. and Gart, J. J. (1971). "Small sample performance of some estimators of the truncated binomial distribution". *J. Amer. Statist. Assoc.*, **66**, 169–77.

5.4 Two parameter Gamma distribution (Ga.D)

5.4.1 Introduction

This distribution has many applications and in particular has been used as a model for precipitation amounts per period (week, month, year) over periods of records at various locations (Thom, 1957, 1958, 1968; Mooley, 1973; Crutcher, Barger and McKay, 1973). In addition, Thom and Vestal (1968) fitted the model (by approximate ml methods) for monthly rainfall at 122 locations in the States (it turned out that for these data the shape parameter is generally near to unity, whereas the scale parameter is frequently either between one and two or small).

5.4.2 Basic formulae

In the estimation of ρ, a from the gamma density

$$f(x; \rho, a) = e^{-x/a}(x/a)^{\rho-1}/a\Gamma(\rho), \quad x > 0 \tag{5.18}$$
$$= 0 \qquad\qquad x < 0$$
$$(a, \rho > 0)$$

the ml estimators are solutions of

$$\ln \hat{\rho} - \psi(\hat{\rho}) = \ln(A/G) \quad (=y) \tag{5.19a}$$
$$\hat{\rho}\hat{a} = A, \tag{5.19b}$$

where A, G are the sample arithmetic and geometric means and $\psi(x) = d \ln \Gamma(x)/dx$. Approximate solutions for $\hat{\rho}$ are $\rho^* = 1/2y$, or better

$$\tilde{\rho} = (1 + \sqrt{(1 + 4y/3)})/4y \qquad \tilde{a} = 3A\{\sqrt{(1 + 4y/3)} - 1\},$$

as studied by Thom (1957, 1958, 1968). Greenwood and Durand (1960) introduced the rational fraction approximations

$$\tilde{\rho} = (0 \cdot 5000876 + 0 \cdot 1648852\, y - 0 \cdot 0544272\, y^2)/y \quad 0 \leq y \leq 0 \cdot 5772,$$

$$\tilde{\rho} = \frac{8 \cdot 898919 + 9 \cdot 059950\, y + 0 \cdot 9775373\, y^2}{y(17 \cdot 79728 + 11 \cdot 968477\, y + y^2)} \quad 0 \cdot 5772 < y \leq 17$$

with errors (in comparison to ml estimators) of 0·008 percent and 0·0054 percent for the two ranges. For $\tilde{a}$, use $\tilde{a}\tilde{\rho} = A$. These are useful for simulation studies.

5.4.3 Expansions for the ml estimators, N large

Using (5.19a), a bivariate Taylor expansion for $\hat{\rho}$ may be set up in the arguments $x = A - \mathrm{E}\,A,\, y = \ln G - \mathrm{E} \ln G$; these are chosen because the joint cumulants of x, y can be found in closed form (Bowman and Shenton, 1968) and in fact

$$\begin{cases} \kappa_{r,s} = \rho(r-1)!\, a^r/N^{r-1}, & s = 0, r \geq 2; \\ \phantom{\kappa_{r,s}} = (r-1)!\, a^r/N^r, & s = 1, r \geq 1; \\ \phantom{\kappa_{r,s}} = \psi_{s-1}(\rho)/N^{s-1}, & r = 0, s \geq 1;\quad (\psi_s(\rho) = d^s\psi/d\rho^s) \\ \phantom{\kappa_{r,s}} = 0, & \text{otherwise.} \end{cases} \tag{5.20}$$

The derivatives of $\hat{\rho}$ with respect to x and y are found from (5.19a) using Faà di Bruno's (1876) formula for the derivative of a function of a function. Inserting the joint moments of x, y in the expectation of $\hat{\rho} - \rho$ and its powers, and filtering out coefficients of N, the sample size, coefficients of powers of N^{-1} as far as N^{-6} have been found for the first four moments. Illustrations are given in Table 5.11.

Table 5.11
Gamma distribution: Bias and variance of $\hat{\rho}$

	ρ	N^{-1}	N^{-2}	N^{-3}	N^{-4}	N^{-5}	N^{-6}
$E(\hat{\rho}-\rho)/\rho$	0·1	1·68	4·11	9·23	14·3	62	578
	0·5	2·17	6·00	17·8	54·4	164	479
	1·0	2·46	7·24	21·8	65·4	195	587
	5·0	2·87	8·61	25·8	77·5	232	697
	25·0	2·97	8·92	26·8	80·3	241	723
$\text{Var}(\hat{\rho}/\rho)$	0·1	1·09	7·24	37	176	807	3697
	0·5	1·36	12·4	85	517	2952	16079
	1·0	1·55	15·8	114	715	4105	22416
	5·0	1·88	20·5	152	959	5540	30379
	25·0	1·97	21·7	162	1020	5897	32352

It will be seen that the bias series for $\hat{\rho}$ consists of positive increasing terms, and for ρ large the ratio of successive terms is nearly 3; similarly the variance terms have common ratio nearly 5 for ρ large. In the same way this property projects for the higher moments. Making use of it, we introduced approximations which extrapolate each series from the last computed coefficient in the form of its asymptotic. For example, our tabulations use the expressions

$$E(\hat{\rho}-\rho)/\rho \sim \sum_{s=1}^{5} \rho_s^{(1)}/N^s + \rho_6^{(1)}/N^5(N-3) \quad (N \geq 4) \quad (5.21)$$

$$\text{Var}\left(\frac{\hat{\rho}}{\rho}\right) \sim \sum_{s=1}^{5} \frac{\rho_s^{(2)}}{N^s} + \frac{\rho_6^{(2)}}{16433}\left[\frac{N^2}{(N-3)^2(N-5)} - \frac{1}{N} - \frac{11}{N^2} - \frac{82}{N^3} - \frac{518}{N^4} - \frac{2995}{N^5}\right]$$
$$(N \geq 6) \quad (5.22)$$

$$\mu_3\left(\frac{\hat{\rho}}{\rho}\right) \sim \sum_{s=2}^{5} \frac{\rho_s^{(3)}}{N^s} + \frac{\rho_6^{(3)}}{25174}$$

$$\times \left[\frac{N^3}{(N-3)^3(N-5)(N-7)} - \frac{1}{N^2} - \frac{21}{N^3} - \frac{271}{N^4} - \frac{2787}{N^5}\right]$$
$$(N \geq 8) \quad (5.23)$$

$$\mu_4\left(\frac{\hat{\rho}}{\rho}\right) \sim \sum_{s=2}^{5} \frac{\rho_s^{(4)}}{N^s} + \frac{\rho_6^{(4)}}{208990}\left[\frac{N^4(N+9)}{(N-3)^4(N-5)(N-7)(N-9)} - \gamma(N)\right]$$
$$(N \geq 10) \quad (5.24)$$

where $\gamma(N) = \frac{1}{N^2} + \frac{42}{N^3} + \frac{937}{N^4} + \frac{15336}{N^5}$.

In each of these, $\rho_s^{(r)}$ refers to the coefficient of N^{-s} in the rth moment of $\hat{\rho}/\rho$. Comparisons of these assessments with other approaches such as asymptotic series in powers of ρ, and Monte Carlo studies have shown that they are very likely subject to less than 5 percent error for $\rho > 0.2$ approximately. Illustrations of the skewness and kurtosis of the distribution of $\hat{\rho}$ are given in Table 5.12.

Table 5.12
Ga.D.: Skewness and kurtosis of $\hat{\rho}$

ρ		Sample size, N						
		14	16	18	20	25	50	100
0.1	$\sqrt{\beta_1}$	1.62	1.42	1.29	1.18	1.01	0.66	0.45
	β_2	10.33	7.98	6.79	6.07	5.11	3.85	3.39
0.5	$\sqrt{\beta_1}$	2.20	1.89	1.68	1.52	1.27	0.80	0.54
	β_2	15.75	11.60	9.44	8.14	6.40	4.26	3.56
1.0	$\sqrt{\beta_1}$	2.35	2.02	1.79	1.63	1.36	0.85	0.57
	β_2	17.16	12.62	10.25	8.79	6.84	4.42	3.63
5.0	$\sqrt{\beta_1}$	2.42	2.08	1.85	1.68	1.40	0.88	0.59
	β_2	17.74	13.09	10.63	9.12	7.08	4.51	3.67
25.0	$\sqrt{\beta_1}$	2.42	2.09	1.85	1.69	1.41	0.88	0.59
	β_2	17.75	13.10	10.64	9.13	7.08	4.52	3.67

Whereas the moment series for $\hat{\rho}$ consist of positive increasing coefficients, those for $\hat{a}$ are variable in sign and do not increase numerically in a regular form; in fact, for ρ large the coefficients for the bias decrease numerically, at least as far as the first few terms are concerned. Sample sizes as small as 3 are sufficient to damp out higher-order terms (see Table 5.13).

It is possible, using an integral representation for the psi function, to prove that $2\hat{\rho}y > 1$ for $\hat{\rho} > 0$, $y > 0$. Now use the moment-generating function of y to derive an expression for $E\, y^{-s}$. From this representation one can prove that mean, variance, third, fourth, ..., moments of $\hat{\rho}$ only exist for samples larger than 3, 5, 7, 9, ..., respectively. For $\hat{a}$, use the fact that $A/\hat{\rho}$ is majorized by $2Ay$ to show that all moments exist for $N \geq 2$ ($N = 1$ being trivial).

Some examples of the skewness and kurtosis of $\hat{a}$ are given in Table 5.14. It will be observed that $\sqrt{\beta_1}$ and β_2 both decrease as ρ increases, whereas for $\hat{\rho}$ the opposite is true.

5.4.4 Expansions for ρ large

Laurent type series for $\hat{\rho}$ may be developed from (5.19a) in terms of y; for example, $\hat{\rho} \sim a_{-1}/y + a_0 + a_1 y + \ldots$, and $\hat{\rho}^s$ has $1/y^s$ as its term of lowest order. Expressions for the first four moments of $\hat{\rho}$ and $\hat{a}$ have been set up (Shenton and Bowman, 1972), using terms as far as ρ^{-20}, each series being

Table 5.13
Ga.D.: Moments of $\hat{a}$

	ρ	N^{-1}	N^{-2}	N^{-3}	N^{-4}	N^{-5}	N^{-6}
	0·2	−0·664	−0·552	−0·042	1·259	0·070	−14·15
	0·5	−0·808	−0·278	0·009	0·184	−0·390	−0·959
$E(\hat{a}-a)/a$	1·0	−0·913	−0·114	0·065	−0·120	0·114	0·823
	5·0	−0·995	−0·001	−0·005	0·002	−0·061	0·999
	10·0	−0·999	−0·000	−0·001	0·000	0·000	0·000
	0·1	11·09	−1·39	−22·5	−8·47	227·8	263·4
	0·5	3·36	−1·29	−2·23	−0·81	1·83	3·62
$Var(\hat{a}/a)$	1·0	2·55	−1·49	−1·26	0·35	0·19	−4·55
	5·0	2·08	−1·95	−0·09	−0·05	0·04	−0·05
	25·0	2·01	−2·00	−0·01	−0·00	0·00	0·00

Table 5.14
Ga.D.: Skewness and kurtosis of $\hat{a}$

N ρ		3	4	6	8	10	20	50	100
0·1	$\sqrt{\beta_1}$	4·89	4·60	3·60	2·99	2·60	1·73	1·05	0·73
	β_2	35·59	37·27	24·97	18·08	14·31	7·91	4·79	3·87
0·5	$\sqrt{\beta_1}$	3·44	2·77	2·10	1·75	1·53	1·04	0·64	0·45
	β_2	24·21	16·47	10·59	8·24	6·99	4·81	3·68	3·33
1·0	$\sqrt{\beta_1}$	2·92	2·30	1·75	1·46	1·28	0·87	0·54	0·38
	β_2	17·32	12·01	8·14	6·58	5·74	4·26	3·48	3·24
5·0	$\sqrt{\beta_1}$	2·17	1·76	1·35	1·14	1·00	0·69	0·43	0·30
	β_2	10·39	7·85	5·87	5·03	4·57	3·74	3·28	3·14
50	$\sqrt{\beta_1}$	2·01	1·64	1·27	1·08	0·95	0·65	0·41	0·29
	β_2	9·11	7·07	5·44	7·74	4·35	3·64	3·25	3·12

truncated at the smallest numerical entry for given N and ρ.

Illustrations of ρ-series are:

Samples of 12

$$Var(\hat{\rho}/\rho) \sim 0.508 - \frac{0.183}{\rho} + \frac{0.045}{\rho^2} + \frac{0.008}{\rho^3} - \frac{0.009}{\rho^4} - \frac{0.010}{\rho^5}$$

$$+ \frac{0.017}{\rho^6} + \frac{0.038}{\rho^7} - \frac{0.093}{\rho^8} - \frac{0.324}{\rho^9} + \frac{1.124}{\rho^{10}} \quad (5.25)$$

$$\beta_2(\hat{\rho}) \sim 29\cdot 400 - \frac{1\cdot 050}{\rho^2} - \frac{0\cdot 880}{\rho^3} + \frac{0\cdot 504}{\rho^4} + \frac{1\cdot 144}{\rho^5} - \frac{1\cdot 254}{\rho^6}$$
$$- \frac{4\cdot 251}{\rho^7} + \frac{7\cdot 848}{\rho^8} + \frac{35\cdot 625}{\rho^9} - \frac{100\cdot 179}{\rho^{10}}. \quad (5.26)$$

5.4.5 Validation of the moments

The two sets of series for the moments in descending powers of N and ρ can be used comparatively for intersecting regions of validity of the parameter space. Four moments of $\hat{\rho}$ in descending powers of N necessitate $\rho \geqslant 0\cdot 1$ and $N \geqslant 14$ approximately for validity of the series; for descending ρ the requirement is $\rho \geqslant 0\cdot 2$, $N > 12$ approximately.

For the moments of $\hat{a}$, the restrictions on sample size are less severe; $N \geqslant 3$ and $\rho \geqslant 0\cdot 1$ are acceptable.

Comparisons of the moments of $\hat{\rho}$ and $\hat{a}$ evaluated by the two asymptotic approaches have been made for $\rho = 1, 1\cdot 5, 2\cdot 0, 2\cdot 5, 5$ and $N = 12, 15, 20, 25, 35, 50$ and the numerical difference is rarely in excess of 2 percent, which is a satisfactory outcome. For example, for $N = 12$, $\rho = 1$, the comparisons for $E(\hat{\rho}/\rho)$, $\text{Var}(\hat{\rho}/\rho)$, $\sqrt{\beta_1(\hat{\rho})}$, $\beta_2(\hat{\rho})$ are (series asymptotic in N given first) $1\cdot 2715$ ($1\cdot 2723$), $0\cdot 3775$ ($0\cdot 3692$), $2\cdot 903$ ($2\cdot 915$), $27\cdot 97$ ($28\cdot 62$). At $N = 12$, $\rho = 5\cdot 0$, the assessments are $1\cdot 3190$ ($1\cdot 3190$), $0\cdot 4731$ ($0\cdot 4731$), $2\cdot 990$ ($2\cdot 991$), $29\cdot 352$ ($29\cdot 422$).

The comparisons for the moments of $\hat{a}$ are equally satisfactory over a similar parameter and sample grid.

5.4.6 Concluding remarks

Detailed tabulations of the moments of $\hat{\rho}$ and $\hat{a}$ have been given elsewhere (Bowman and Shenton, 1968, 1970; Shenton and Bowman, 1972). From the estimation point of view, almost unbiassed estimators for $\hat{\rho}, \hat{a}$ and their variances have been mentioned in these papers; for small samples ($3 < N < 50$) asymptotic biases and variances can be seriously defective.

For the purposes of making inferences on the parameters ρ and a for given sample sizes, the distribution functions of $\hat{\rho}$ and $\hat{a}$ can be approximated by a four-moment Pearson distribution, followed by the exact percentage points of the latter derived from Pearson and Hartley (1972). To validate these approximate percentage points (or others derived from two-moment approximating distributions) would require an extensive investigation over sample sizes, percentiles (perhaps included within 5 percent and 95 percent) and values of ρ; in this connection it is useful to recall that $\hat{a}/a$ and $\hat{\rho}$ are distributed independently of a, which leads to some economy of effort for tabulation purposes. Even then, the assessments of percentiles can only achieve acceptance by comparison with other approximations (for a related problem, one may consider the extensive studies directed at the percentiles of the skewness and kurtosis statistics in sampling from the normal (see, for example, D'Agostino and Pearson, 1973) for which the sample size and percentile are the only variables).

However, we have tested out a few isolated points with different approximating distributions (Table 5.15); agreement is satisfactory. Further work in this direction would require an efficient fast generator for Gamma variates (including non-integral shape parameter) and a quick method of evaluating $\hat{\rho}$ and $\hat{a}$.

Table 5.15
Ga.D.: Selected percent points of the distributions of $\hat{\rho}$, $\hat{a}$

ρ	N	Variate		Percentile					
				1	5	10	90	95	99
1	50	$\hat{\rho}$	Pearson	0·702	0·779	0·826	1·309	1·409	1·628
			S_U	0·700	0·780	0·826	1·309	1·410	1·629
			Monte Carlo	0·704	0·782	0·828	1·306	1·408	1·636
			Dy	0·701	0·780	0·826	1·309	1·409	1·628
		$\hat{a}/a$	Pearson	0·546	0·649	0·710	1·278	1·381	1·592
			Monte Carlo	0·544	0·649	0·709	1·275	1·375	1·582
2	50	$\hat{\rho}$	Pearson	1·378	1·539	1·637	2·659	2·873	3·342
			S_U	1·375	1·540	1·638	2·659	2·874	3·343
			Monte Carlo	1·387	1·545	1·639	2·659	2·872	3·335
			Dy	1·378	1·539	1·637	2·660	2·873	3·342
		$\hat{a}/a$	Pearson	0·565	0·666	0·725	1·257	1·350	1·539
			Monte Carlo	0·567	0·666	0·727	1·252	1·346	1·528

(i) Pearson refers to the percentiles of the corresponding 4-moment Pearson distribution. (ii) Monte Carlo runs were based on 50000 samples. (iii) S_U refers to Johnson's transformed normal distribution. (iv) Dy refers to percent points derived from the use of the distribution of y.

Another aspect of the estimation problem concerns the interpretation of results when ρ and a (or $c = 1/a$) are unknown; for example, the bias of $\hat{\rho}$ involves ρ which is not available. We have constructed, using asymptotic series for the mle $\hat{\rho}, \hat{a}, \hat{c}$, the following *almost* unbiased estimators, on the assumption that ρ is large.

$$\hat{\hat{\rho}} = \frac{(N-3)}{N}\hat{\rho} + \frac{2}{3N} - \frac{1}{9N\hat{\rho}} - \frac{7}{54(N+3)N\hat{\rho}^2} - \frac{(26N^2 - 55N - 169)}{810(N+3)(N+5)N\hat{\rho}^3}$$

$$v_\rho = \frac{2\hat{\rho}^2}{N-3} - \frac{2\hat{\rho}}{3(N-3)} + \frac{2N-8}{9(N-3)^2} + \frac{8N^2 + 10N - 112}{135(N-3)^2(N+3)\hat{\rho}}$$

$$\hat{\hat{a}} = A\left\{\frac{2Ny}{N-1} - \frac{2Ny^2}{3(N-1)} + \frac{4N(N+1)y^3}{9(N-1)(N+3)} - \frac{2N(7N^2 + 60N + 7)y^4}{135(N-1)(N+3)(N+5)}\right\}$$

$$\hat{v}_a = A^2 \left\{ \frac{8y^2}{N+1} - \frac{8(N+7)y^3}{3(N+1)(N+3)} + \frac{16(N+4)y^4}{3(N+1)(N+5)} \right\}$$

$$\hat{\rho}_y = \frac{N-3}{2Ny} + \frac{N+1}{6N} - \frac{(N+1)y}{18N} - \frac{(4N^2 - 10N + 4)}{135N(N+3)} y^2$$

$$\hat{v}_\rho(y) = \frac{1}{2(N-3)y^2} + \frac{N-7}{18(N-3)^2} + \frac{4(2N^2 + 5N - 38)}{135(N-3)^2(N+3)} y$$

$$\hat{c} = A^{-1} \left\{ \frac{N-3}{2Ny} + \frac{N-5}{6N} - \frac{(N+1)y}{18N} - \frac{2(2N^2 - 5N + 2)y^2}{135N(N+3)} \right\}$$

$$\hat{v}_c = A^{-2} \left\{ \frac{1}{2(N-3)y^2} + \frac{1}{2(N-3)y} + \frac{2N-5}{9(N-3)^2} \right.$$

$$\left. + \frac{(N^3 + 99N^2 - 29N + 169)}{270(N-1)(N-3)^2(N+3)} y \right\}.$$

(A is the sample arithmetic mean.)

For these, after some algebraic simplification, we find the following mean values:

$$E\hat{\rho} = \rho + O(1/\rho^4)$$
$$E v_\rho = \text{Var } \hat{\rho} + O(1/\rho^2)$$
$$E\hat{a} = a + O(1/\rho^4)$$
$$E\hat{v}_a = \text{Var } \hat{a} + O(1/\rho^3)$$
$$E\hat{\rho}_y = \rho + O(1/\rho^3)$$
$$E\hat{v}_\rho(y) = \text{Var } \hat{\rho} + O(1/\rho^2)$$
$$E\hat{c} = c + O(1/\rho^3)$$
$$E\hat{v}_c = \text{Var } \hat{c} + O(1/\rho^2).$$

Further investigation of these results, perhaps by Monte Carlo simulation, would be of interest.

In connection with unbiasedness and minimum variance, and the search for most efficient estimation procedures, the reader is referred to an interesting study by Blischke (1971).

REFERENCES

1. Blischke, W. R. (1971). "Further results on estimation of the parameters of the Pearson Type III distribution". Report ARL 71-0063, U.S. Department of Commerce, Springfield, Virginia.
2. Bowman, K. O. and Shenton, L. R. (1968). "Properties of estimators for the gamma distribution". CTC-1 Report, Union Carbide Corporation, Nuclear Division, Oak Ridge, Tennessee.

3. Bowman, K. O. and Shenton, L. R. (1970). "Small sample properties of estimators for the gamma distribution". Report CTC-28, Union Carbide Corporation, Nuclear Division, Oak Ridge, Tennessee.
4. D'Agostino, R. and Pearson, E. S. (1973). "Tests for departure from normality. Empirical results for the distributions of b_2 and $\sqrt{b_1}$". *Biometrika*, **60**, 3, 613–22.
5. Crutcher, H. L., Barger, G. L. and McKay, G. F. (1973). "A note on a gamma distribution computer program and graphic paper". *NOAA Technical Report EDS11*, U.S. Dept. of Commerce, pp. 1–38.
6. Faà di Bruno (1876). *Théorie des Formes Binaires*, Torino, Librairie Brero Publ.
7. Greenwood, J. A. and Durand, D. (1960). "Aids for fitting the gamma distribution by maximum likelihood". *Technometrics*, **2**, 55–65.
8. Mooley, Diwakar (1973). "An estimate of the distribution and stability period of the parameters of the gamma probability model applied to monthly rainfall over Southeast Asia during the summer monsoon". *Monthly Weather Rev.*, **101**, No. 12, pp. 884–90.
9. Pearson, E. S. and Hartley, H. O. (1972). *Biometrika Tables for Statisticians*, Vol. 2, Cambridge University Press.
10. Shenton, L. R. and Bowman, K. O. (1972). "Further remarks on maximum likelihood estimators for the gamma distribution". *Technometrics*, **14**, 725–33.
11. Thom, H. C. S. (1957). "A statistical method of evaluating augmentation of precipitation by cloud seeding". *Final Report, U.S. Advisory Committee on Weather Control*, Vol. 2, Washington, D.C., pp. 5–25.
12. Thom, H. C. S. (1958). "A note on the gamma distribution". *Monthly Weather Rev.*, **86**, No. 4, 117–22.
13. Thom, H. C. S. (1968). "Approximate convolution of the gamma and mixed gamma distribution". *Monthly Weather Rev.*, **96**, No. 12, 883–6.
14. Thom, H. C. S. and Vestal, I. B. (1968). "Quantiles of monthly precipitation for selected stations in the contiguous United States". *ESSA Technical Report*, EDS 6.

6 SUMMARY AND CONCLUSIONS

6.1 The subject of estimation is vast in its scope, and here we have confined our attention to maximum likelihood estimators. Other methods are available, and circumstances, aims, and individual choice must play their roles. We certainly do not recommend ml, least squares, or any other procedure as the only approach worth considering.

6.2   The method of moments

The method of moments is simple to use and of wide generality. To estimate $\theta_1, \theta_2, \ldots, \theta_h$ by moments we choose a (mathematically) consistent set of equations based on h (statistically) consistent moment estimators. The moments may be either noncentral or central, according to convenience; in the discrete case, factorial moments or factorial cumulants may be advantageous from the viewpoint of simplicity, but their sampling moments are more complicated than those for noncentral or central moments.

Moments need not be confined to expectations of polynomials in the variate; quite general functions can be used. Thus, although it is not recommended, $\sum_1^n \ln |x_j|$ could be used to estimate a function of σ in sampling from $N(0, \sigma^2)$; more usefully, the expectation of $(\ln x)^s$, $s = 1, 2, \ldots$, can be evaluated in sampling from a generalized gamma distribution, with density

$$f(x) = kx^{c\alpha - 1} \exp - (x/\beta)^c, \quad x > 0$$
$$ = 0, \text{ otherwise} \qquad (\alpha, c, \beta > 0),$$

so that in this case $\Sigma (\ln x_j)^s$ would provide perfectly acceptable estimators (Johnson and Kotz, 1970, pp. 197–8).

Similarly, for the three-parameter gamma (**4.5**) the generalized moment estimators derived from $Ex = s + a\rho$, $E(x - s)^{-1} = 1/a(\rho - 1)$, and $E \ln(x - s) = \psi(\rho) + \ln(a)$ are in fact mle (Bowman and Shenton, 1969; ref. 1 in **4.5**).

6.3   Moment estimators related to the likelihood function

Let $P(x; \theta)$ be the probability of the variate x depending on the h-parameters $\theta_1, \theta_2, \ldots, \theta_h$ (we confine our attention to the continuous case here, since this avoids the use of Stieltjes integrals). Assume that $P(x; \theta)$ possesses first derivatives with respect to θ for $\theta \in \Theta$, and that the moments

$\mu'_r = E\,x^r$ and their first derivatives exist and are finite. Let $\{q_r(x)\}$ be the orthogonal system of polynomials associated with $P(x;\theta)$, and define

$$q_r(x) = \sum_{j=0}^{r} a_{rj} x^j \quad (a_{rr} \neq 0,\ r = 0, 1, \ldots),$$

where

$$\int q_r^2(x) P(x;\theta)\,dx = \phi_r \quad (>0),$$
$$\int q_r(x) q_s(x) P(x;\theta)\,dx = 0 \quad (r \neq s).$$

Let the formal Fourier expansions of the derivatives of P be

$$\frac{\partial P(x;\theta)}{\partial \theta_j} = P(x;\theta) \sum_{r=0}^{\infty} A_{jr} q_r(x) \quad (j = 1, 2, \ldots, h). \tag{6.1}$$

For the partial sums of the series in (6.1) write

$$A_j^{(r)}(x) = \sum_{k=0}^{r} A_{jk} q_k(x),$$

where $\quad A_{jk}\phi_k = \int q_k(x) \dfrac{\partial P(x;\theta)}{\partial \theta_j} dx = -\int \dfrac{\partial q_k(x)}{\partial \theta_j} P(x;\theta) dx,$

if the range is independent of θ, which is assumed to be the case. $A_j^{(r)}(x)$ is now a polynomial approximation to $\partial \log P / \partial \theta_j$, so that the parameter θ_j may be estimated from

$$A_j^{(r)} = \sum_{t=1}^{N} A_j^{(r)}(x_t)/N = 0 \quad (j = 1, \ldots, h), \tag{6.2}$$

if these equations are mathematically consistent and $r \geqslant h$ (actually r does not have to be the same for each j; however, at least one r must at least equal h); doubtless, under certain regularity conditions the moment estimators $(\theta_1^*, \theta_2^*, \ldots, \theta_h^*)$ converge to the mle $\hat{\theta}$. It may be shown that an alternative expression to (6.2) is

$$\begin{vmatrix} 0 & m'_0 & m'_1 & \cdots & m'_r \\ 0 & \mu'_0 & \mu'_1 & \cdots & \mu'_r \\ \dfrac{\partial \mu'_1}{\partial \theta_j} & \mu'_1 & \mu'_2 & \cdots & \mu'_{r+1} \\ \vdots & \vdots & \vdots & & \vdots \\ \dfrac{\partial \mu'_r}{\partial \theta_j} & \mu'_r & \mu'_{r+1} & \cdots & \mu'_{2r} \end{vmatrix} = 0 \quad \begin{array}{c} (j = 1, \ldots, h) \\ (r \geqslant h) \end{array} \tag{6.3}$$

where m'_r is the sample noncentral moment and $E\,m'_r = \mu'_r$.

Example 6.1 Show that the moment estimators of θ and v in sampling from $N(\theta, v)$ are $\theta^* = m'_1$, $v^* = m'_2 - m'^2_1$.

It may also be shown in the single-parameter case that if
$$A_s \phi_s = - \int P(x;\theta) \frac{\partial q_s(x)}{\partial \theta} dx, \qquad (6.4)$$
and θ_r^* is a solution to (6.2), then the asymptotic variance is given by
$$(N \operatorname{Var} \theta_r^*)^{-1} = A_1^2 \phi_1 + A_2^2 \phi_2 + \ldots + A_r^2 \phi_r, \qquad (6.5)$$
with asymptotic efficiency
$$E_f(\theta_r^*) = \sum_{s=1}^{r} A_s^2 \phi_s / \sum_{s=1}^{\infty} A_s^2 \phi_s; \qquad (6.6)$$
moreover, as more sample moments are used,
$$\operatorname{Var} \theta_1^* \geqslant \operatorname{Var} \theta_2^* \geqslant \ldots \geqslant \operatorname{Var} \theta_r^* \ldots,$$
which suggests that the limiting variance is the asymptotic variance of the mle.

For further discussion see Shenton (1958, 1959) and Shenton and Wallington (1962).

Example 6.2 Consider estimating θ, v in sampling from $N(\theta, v)$. In this case,
$$\frac{1}{f} \frac{\partial f}{\partial \theta} = \frac{(x-\theta)}{v}, \quad \frac{1}{f} \frac{\partial f}{\partial v} = -\frac{1}{2v} + \frac{(x-\theta)^2}{2v^2},$$
so that θ^*, v^* coincide with the mle. Similarly, the estimators of θ, σ from $N(\theta, \sigma^2)$ would be the usual mle.

Example 6.3 Show that if the logarithmic derivatives of the probability density (or probability function) with respect to the parameters are polynomials in the variate, then the generalized moment estimators are in general mle.

Example 6.4 Consider generalized moment estimators for the two-parameter gamma density (5.4). Show that
$$E \exp(\alpha x - a\rho\alpha) = \exp\{-a\rho\alpha - \rho \ln(1 - a\alpha)\} \quad (a\alpha < 1).$$
Deduce that
$$E(\ln(x/a) - \psi(\rho)) = 0,$$
$$E(\ln(x/a) - \psi(\rho))(x - a\rho) = 0,$$
$$E(\ln(x/a) - \psi(\rho))(x - a\rho)^{s+1} = a^s(s-1)! \mu_0 + a^{s-1}\binom{s}{1}(s-2)! \mu_1$$
$$+ a^{s-2}\binom{s}{2}(s-3)! \mu_2 + \ldots + a\binom{s}{s-1}\mu_{s-1}, \quad (s = 1, 2, \ldots),$$

where $\psi(\rho)$ is the psi function and $\{\mu_s\}$ are central moments of the distribution ($\mu_0 = 1$, $\mu_1 = 0$, $\mu_2 = a\rho$, etc). Noting that

$$\frac{1}{f}\frac{\partial f}{\partial a} = \frac{x}{a^2} - \frac{\rho}{a}$$

$$\frac{1}{f}\frac{\partial f}{\partial \rho} = \ln(x/a) - \psi(\rho),$$

show that the "simplest" generalized moment estimators are the usual ones, namely,

$$a^*\rho^* = m_1', \quad a^{*2}\rho^* = m_2.$$

Show that the next pair are

$$a^*\rho^* = m_1',$$
$$3(\rho^* + 2)(m_2 - a^*m_1') + a^*(3\rho^* - 4)(m_3 - 6a^*m_2 + 4a^{*2}m_1') = 0.$$

Before leaving moment estimators, note that if an estimation structure is defined by the sum of random variates, then mle may be complicated (see Sclove and Ryzin, 1969). But if, for example,

$$Z = X + Y,$$

where X and Y are independent, with moments to a certain order, then a system of moment estimators may be derived from $E(X + Y)^s$ expressed in terms of the two population moments.

Investigations into the importance of second-order terms for moment estimators have been carried out by Robertson and Fryer (1970) in connection with normal mixtures. Mason (1973) has looked into the part played by N^{-2} terms in the biases and covariances of moment estimators for a four-parameter model which is basically a density consisting of a truncated gamma variate continuous with another gamma variate; i.e., the discontinuity in the density is a parameter to be estimated.

6.4 Mixed moment estimators

Mixed estimators can sometimes be used advantageously. For example, in estimating s, a, ρ from the three-parameter gamma (4.5) one could use $x_{\min}$ for s, and determine a and ρ by conditional maximum likelihood. Again, if a density has a discontinuity at an unknown point $x = b$, then the ml approach becomes problematical; however, trial values of b could be tested to locate the likelihood maximum over the rest of the parameters (assuming that a maximum exists).

6.5 Estimators derived from measures of discrepancy

There are numerous measures of discrepancy whose minimization may lead to sets of estimators, particularly for discretized data or for grouped data. Rao (see ref. 43 in Chapter 1, above) lists

(i) $\chi^2 = \sum \dfrac{(n_i - Np_i(\theta))^2}{Np_i(\theta)}$

and modified χ^2 in which the denominator in χ^2 is replaced by n_i;

(ii) Hellinger distance:
$$HD = \arccos \sqrt{\sum (n_i/N) p_i(\theta)};$$

(iii) Haldane's discrepancy:
$$D_k = \frac{(n+k)!}{n!} \sum \frac{n_i! p_i^{k+1}(\theta)}{(n_i+k)!} \quad (k \neq -1)$$

$$D_{-1} = -\frac{1}{N} \sum n_i \ln p_i(\theta);$$

(iv) Kullback–Leibler separator:
$$KLS = \sum p_i(\theta) \ln \left(\frac{p_i(\theta)}{n_i/N} \right).$$

There may well be other procedures which would generate estimators with good or "best" properties.

6.6 Moment estimators and numerical evaluation

Moment estimators are quite often easily evaluated numerically and can be used as initial values in iteration schemes for ml estimates. This is the case, for example, for the negative binomial, Neyman Type A and two- and three-parameter gamma distributions (see sections 4.2 to 4.6). Thus, for the Neyman Type A with parameters θ_1, θ_2 (4.3), in terms of sample moments, the moment estimators are

$$\theta_2^* = m_2/m_1 - 1, \quad \theta_1^* = m_1^2/(m_2 - m_1).$$

For ml, we have (Shenton, 1949)

$$\hat{\theta}_1 \hat{\theta}_2 = \frac{1}{N} \sum' n_x (x+1) \hat{P}_{x+1}/\hat{P}_x$$
$$= m_1. \tag{6.7}$$

Defining $F(\theta_2) = \sum' n_x(x+1) P_{x+1}/P_x - Nm_1$, in which P_x, P_{x+1} have θ_1 replaced by m_1/θ_2, an iterative scheme for θ_2 can be set up with initial value θ_2^*. Thus the increment is given by

$$\delta \theta_2 = -F(\theta_2)/F^{(1)}(\theta_2), \tag{6.8}$$

where the superscript refers to a derivative, and

$$F^{(1)}(\theta_2) = \frac{1}{\theta_2} \sum{}' n_x(x+1)\frac{P_{x+1}}{P_x} - \frac{(\theta_2+1)}{\theta_2^2}$$
$$\times \sum{}' \left\{ (x+1)(x+2)\frac{P_{x+2}}{P_x} - (x+1)^2 \frac{P_{x+1}}{P_x^2} \right\}, \quad (6.9)$$

the terms in P_x being computed with the value $\hat{\theta}_2$ of θ_2 at the previous iterate, and $\theta_1 = m_1/\hat{\theta}_2$.

Of course, any other estimator which is easily calculated can be used as initializer, and a general iterative scheme can be based on the likelihood function (see Kendall and Stuart, ref. 34 in Chapter 1, pp. 51–4). Thus, in the single-parameter case, an iterative scheme is

$$\hat{\theta}_{m+1} = \hat{\theta}_m + \left(\frac{\partial \ln L}{\partial \theta}\right)_{\theta = \hat{\theta}_m} \text{Var } \hat{\theta}, \quad m = 0, 1, \ldots, \quad (6.10)$$

where $\hat{\theta}_0 = t$, the simple estimator, and Var $\hat{\theta}$ the variance of the mle; this may depend on θ, in which case $\hat{\theta}_m$ is used.

The generalization to the multiparameter case is fairly obvious. Iterative schemes, which will in general be programmed for a digital computer, may turn out to be numerically unstable and sensitive to the initiating vector, especially for small samples and numerous parameters. If many iterates are required (say 50 or more), some attention should be given to the question of numerical accuracy and various sources of accumulation of error (single-precision arithmetic, for example, may be inadequate). Graphical approaches to arrive at initializers should not be ruled out, and final estimates should be checked, where possible, to determine if the likelihood has been maximized.

6.7 Conclusions

ML has enjoyed a long period of usefulness, though controversies have, from time to time, apparently engulfed it. Many of these were concentrated on local aspects and unwittingly ignored the global.

For example, it is strange to recall how frequently in the early development of the subject the concepts of largeness and asymptotics were taken for granted. How many textbooks in recent years, which do justice to the subject, explain the meaning of an asymptotic result? And yet the majority of the results are only true for some unattainable sample size. Even so, this criticism is not altogether fair or valid; for there may well be some sample size N^* for which consistent solutions start to appear, for which variances are not too discrepant from the minimum possible, and for which the departure from normality (measured in some sense) is not important for the purpose at hand. The controversy between Karl Pearson and R. A. Fisher some fifty years ago surrounding efficient statistics was largely irrelevant and now appears particularly incomprehensible, since both these writers were exceptionally capable numerical analysts for their era, yet each of them seemed completely unaware of the fatal flaw involved in ignoring the asymptotic nature of the basic results.

Again, the use of special examples has played an important part in the development of the appreciation of estimator properties. Superefficient estimators, structures for which there is no consistent estimator, and so on, have been discovered, and have been used by some writers (usually not the originators) to discredit the underlying procedures. Far from detracting from the theories, these special malfunctions of the theory ultimately enrich the total concept.

Research is still needed to discover how small-sample estimators perform when various good properties are hoped for. Is the aim a set of estimators nearly normally distributed (as measured, say, by the nearness to the 95 percent and 5 percent points)? Is unbiasedness or biasedness an overriding requirement? (Bias would be of importance if a packaged weight of an industrial product was involved.) Modern computers, along with the needed elementary programming knowledge which most science students nowadays possess, can provide a large number of interesting and creative endeavours associated with estimation problems. For example, the programming of the higher-order terms for the first four moments of mle in the single-parameter case (see formulae (2.30), (2.31), (2.33) and (2.34)) would form a starting point. This would provide a real feeling for the essence of the estimation problem. Again, even a modest pocket programmable electronic computer can provide a facility for ingenuity and insight into means and variances, etc. and their asymptotic counterparts. Further, Monte Carlo simulation is now well within the reach of most students, and it can enlarge the insight into statistical models and call attention to the reliability of associated parameter estimates.

Lastly, the question of the validity and usefulness of higher-order terms seems to us to raise several new problems, and these problems occur in many situations other than those described here; as a corollary, the failure to analyse the validity and usefulness of dominant asymptotic formulae in estimation theory (and elsewhere in statistical theory) should cause concern. It is no longer entirely a matter of correct algebraic manipulation. The new problems are at the interface of statistical theory and analysis and computer science. How do we satisfy ourselves that fairly massive computations committed to computers and resulting in coefficients of N^{-2} to N^{-9} terms in moments are correct to a specified number of digits? Errors can arise from slips in the theory, from programmers, and occasionally from machine malfunctions, to say nothing of rounding-off problems and problems relating to orders of magnitude of terms and sums of terms.

It is sometimes said that many of the easy research problems have already been solved. Be that as it may, digital facilities are providing many difficult and complex problems possibly undreamt of a decade or so ago.

REFERENCES

1. Johnson, N.L. and Kotz, S. (1970). *Distributions in Statistics – Continuous Univariate Distributions–I*. Houghton Mifflin, Boston.
2. Mason, T.J. (1973). *A Statistical Model for Induced Disease*. Dissertation in part fulfilment of the Ph.D. Degree, Univ. of Georgia.
3. Robertson, C.A. and Fryer, J.G. (1970). "The bias and accuracy of moment estimators". *Biometrika*, **57**, 57–65.
4. Sclove, S.L. and Ryzin, J. van (1969). "Estimating the parameters of a convolution". *J. Roy. Statist. Soc. (B)*, **31**, No. 1, 181–91.
5. Shenton, L.R. (1949). "On the efficiency of the method of moments and Neyman's Type A distribution". *Biometrika*, **36**, 450–4.
6. Shenton, L.R. (1958). "Moment estimators and maximum likelihood". *Biometrika*, **45**, 411–20.
7. Shenton, L.R. (1959). "The distribution of moment estimators". *Biometrika*, **46**, 296–305.
8. Shenton, L.R. and Wallington, P. (1962). "The bias of moment estimators with an application to the negative binomial distribution". *Biometrika*, **49**, 193–204.

APPENDIX A

A FACTORIAL SERIES (IN N) FOR $E(\hat{\theta} - \theta)^s$

A.1 Introduction

Inspection of the series for $E(\hat{\theta} - \theta)$ given in (2.30a) shows it to have the form

$$\frac{a}{NI^2} + \frac{1}{N^2}\left(-\frac{a}{I^2} + \frac{b}{I^3} + \frac{c}{I^4} + \frac{d}{I^5}\right)$$
$$+ \frac{1}{N^3}\left(\frac{a}{I^2} - \frac{3b}{I^3} + \frac{c'-3c}{I^4} + \frac{d'-3d}{I^5} + \ldots + \frac{g}{I^8}\right) + \ldots, \quad (A.1)$$

I being Fisher's information function.

This suggests a factorial form of expansion

$$\frac{a}{(N+1)I^2} + \frac{1}{(N+1)(N+2)}\frac{b}{I^3} + \frac{c}{I^4} + \frac{d}{I^5} + \ldots \quad (A.2)$$

Closer inspection now shows that with this factorial formulation, the coefficients in a, b, c, d, etc. (each of which involves products of expectations of products of logarithmic derivatives of the probability function) are all positive; thus the terms in $b/I^3 + c/I^4 + d/I^5$ are those with positive coefficients in $A_2^{(1)}$ in (2.30a); similarly the terms multiplying

$$1/\{(N+1)(N+2)(N+3)\}$$

are those in $A_3^{(1)}$ with negative-signed terms deleted.

A further property becomes clear if we analyse the terms partitionally. Define $\mathcal{P}(f|m)$ as a set of terms each consisting of f logarithmic derivative factors (2.29), the assemblage of factors being a partition of m. For example, $\mathcal{P}(2|7)$ consists of terms such as $[4][1^3]$, $[3][2^2]$, $[123][1]$, etc. The structure of the positive terms (not ultimately necessarily positive) in $E(\hat{\theta} - \theta)$ is now seen to be

$$\frac{1}{N+1} \to \frac{\mathcal{P}(1|3)}{I^2}$$

$$\frac{1}{(N+1)(N+2)} \to \frac{\mathcal{P}(1|5)}{I^3}, \frac{\mathcal{P}(2|7)}{I^4}, \frac{\mathcal{P}(3|9)}{I^5}$$

$$\frac{1}{(N+1)(N+2)(N+3)} \to \frac{\mathcal{P}(1|7)}{I^4}, \frac{\mathcal{P}(2|9)}{I^5}, \ldots, \frac{\mathcal{P}(5|15)}{I^8},$$

with an obvious indication that in general

$$\frac{1}{(N+1)(N+2)\ldots(N+s)} \to \frac{\mathcal{P}(1|2s+1)}{I^{s+1}}, \frac{\mathcal{P}(2|2s+3)}{I^{s+2}}, \ldots,$$
$$\frac{\mathcal{P}(2s-1|6s-3)}{I^{3s-1}},$$

where $\to$ means "consists of".

A.2 Restrictions on the partitions

A.2.1 Now consider the set of terms in $E(\hat{\theta} - \theta)$ of order up to and including N^{-s}. For this we have to study the Lagrange expansion

$$\hat{\theta} - \theta = \sum_{1}^{\infty} L_1^r \Theta_r / (r!\, L_2^r), \tag{A.3}$$

where, for example (see Comtet, 1974),

$\Theta_1 = -1,$

$\Theta_2 = -L_3/L_2,$

$\Theta_3 = L_4/L_2 - 3L_3^2/L_2^2,$

$\Theta_4 = -L_5/L_2 + 10L_3 L_4/L_2^2 - 15L_3^3/L_2^3,$

$\Theta_5 = L_6/L_2 - 15L_3 L_5/L_2^2 - 10L_4^2/L_2^2 + 105L_3^2 L_4/L_2^3 - 105L_3^4/L_2^4,$

$\Theta_6 = -L_7/L_2 + 21L_3 L_6/L_2^2 + 35L_4 L_5/L_2^2 - 210L_3^2 L_5/L_2^3$
$\quad - 280L_3 L_4^2/L_2^3 + 1260L_3^3 L_4/L_2^4 - 945L_3^5/L_2^5,$

$\Theta_7 = L_8/L_2 - 28L_3 L_7/L_2^2 - 56L_4 L_6/L_2^2 - 35L_5^2/L_2^2 + 378L_3^2 L_6/L_2^3$
$\quad + 1260L_3 L_4 L_5/L_2^3 + 280L_4^3/L_2^3 - 3150L_3^3 L_5/L_2^4 - 6300L_3^2 L_4^2/L_2^4$
$\quad + 17325L_3^4 L_4/L_2^5 + 10395L_3^6/L_2^6,$

$\Theta_8 = -L_9/L_2 + 36L_3 L_8/L_2^2 + 84L_4 L_7/L_2^2 + 126L_5 L_6/L_2^2$
$\quad - 630L_3^2 L_7/L_2^3 - 2520L_3 L_4 L_6/L_2^3 - 1575L_3 L_5^2/L_2^3$
$\quad - 2100L_4^2 L_5/L_2^3 + 6930L_3^3 L_6/L_2^4 + 34650L_3^2 L_4 L_5/L_2^4$
$\quad + 15400L_3 L_4^3/L_2^4 - 51975L_3^4 L_5/L_2^5 - 138600L_3^3 L_4^2/L_2^5$
$\quad + 270270L_3^5 L_4/L_2^6 - 135135L_3^7/L_2^7.$

The general term is quite straightforward to write down and consists of products of L-terms which are (a) one-part partitions of $r+1$, (b) two-part partitions of $r+3, \ldots$, (h) $(r-1)$ part partitions of $3r-3$ (see for example McMahon, 1894).

To collect all terms in the bias to order N^{-s}, we have only to consider terms in (A.3) from the first power of L_1 to L_1^{2s}; thus, N^{-s} terms arise from $2s$ terms of (A.3). Moreover, the partitions occurring are subject to restrictions which are derived from (A.3). For example, logarithmic derivative terms (ldt's) in the $1/(N+1)$ coefficient are limited to using L_1, L_2, L_3, which occur in the first two terms.

A.2.2 Expansion for $E(L_1/L_2)$

From 2.4 we have

$$E\, e^{\alpha L_1 + \beta L_2} = \{\Sigma\, p_j\, e^{(\alpha \partial_1 + \beta \partial_2) P_j/N}\}^N, \tag{A.4}$$

where $\partial_s P_j \equiv \dfrac{\partial^s}{\partial \theta^s} \ln p_j$. Differentiating,

$$E L_1 e^{\beta L_2} = \{\Sigma\, p_j\, e^{\beta \partial_2 P_j/N}\}^{N-1} \Sigma\, (p_j \partial_1 P_j) e^{\beta \partial_2 P_j/N}$$
$$= (p\, e^{\beta \partial_2 P/N})^{N-1}(p \partial_1 P_1\, e^{\beta \partial_2 P/N}),$$

using summatory notation. From this, formally,

$$E \int_0^{N/I} L_1\, e^{-\beta(-L_2)} d\beta = \int_0^{N/I} \Phi^{N-1}(\beta)(\Phi_1(\beta))\, d\beta, \tag{A.5}$$

where

$$\Phi(\beta) = 1 - \frac{\beta I}{N} + \frac{\beta^2 [2^2]}{2! N^2} + \ldots,$$

$$\Phi_1(\beta) = \Sigma\, \frac{\partial p_j}{\partial \theta} + \Sigma\, \beta(p_j \partial_1 P_j)(\partial_2 P_j)/N$$

$$+ \Sigma\, \beta^2 (p_j \partial_1 P_j)(\partial_2 P_j)^2/(2!\, N^2) + \ldots$$

$$= \beta[12]/N + \beta^2 [12^2]/(2!\, N^2) + \ldots$$

in terms of ldt's.

From (A.5), assuming that $\exp(NL_2/I)$ is of negligible order for large N,

$$E\left(-\frac{L_1}{L_2}\right) \sim \frac{N}{I} \int_0^1 \left(1 - y + \frac{[2^2] y^2}{2!} + \ldots\right)^{N-1} \left\{\frac{[12] y}{I} + \frac{[12^2] y^2}{2 I^2} + \ldots\right\} dy$$

$$= \int_0^1 \sum_1^N (1-y)^{N-s} \Psi_s(y)\, dy, \tag{A.6}$$

where

$$\Psi_s(y) = \frac{N}{1}\binom{N-1}{s-1}\left(\sum_2^\infty \frac{[2^r] y^r}{r!}\right)^{s-1} \sum_1^\infty \frac{[12^r] y^r}{I^r r!}.$$

The first few terms yield

$$E\left(-\frac{L_1}{L_2}\right) \sim \frac{[12]}{M_1 I^2} + \frac{1}{M_2}\left\{\frac{[12^2]}{I^3} + \frac{3[12][2^2]}{I^4}\right\}$$

$$+ \frac{1}{M_3}\left\{\frac{[12^3]}{I^4} + \frac{6[12^2][2^2]}{I^5} + \frac{4[12][2^3]}{I^5} + \frac{15[12][2^2]^2}{I^6}\right\} + \ldots,$$

where $M_s \equiv (N+1)(N+2) \ldots (N+s)$. It will be noted that the M_1^{-1} coefficient is an ldt associated with a partition of 3, the M_2^{-1} coefficient relates to partitions $\mathcal{P}(1|5), \mathcal{P}(2|7)$, and the M_3^{-1} coefficient relates to $\mathcal{P}(1|7), \mathcal{P}(2|9), \mathcal{P}(3|11)$.

A.2.3 Expansion for $E(L_1^3 L_4 / 6 L_2^4)$

For this (and terms of higher degree in (A.3)) we need an expression for the derivative of a function of a multivariate function (see for example Riordan (1946), Good (1961); also "Tables of crude moments expressed in terms of cumulants" (TCM Tables) by Kratky, Reinfelds, Hutcheson and Shenton, Computer Center Report Number 7201, University of Georgia). Starting with

$$E\, e^{\alpha L_1 + \beta L_2 + \gamma L_4} = \{\Sigma\, p_j\, e^{(\alpha \partial_1 + \beta \partial_2 + \gamma \partial_4) P_j / N}\}^N, \tag{A.7a}$$

we need

$$\left. \frac{\partial^4 \Phi^N(\alpha, \beta, \gamma)}{\partial \alpha^3 \partial \gamma} \right|_{\alpha = \gamma = 0}$$

where

$$\Phi(\alpha, \beta, \gamma) = \left\{ p \exp\left(\frac{\alpha}{N} \partial_1 + \frac{\beta}{N} \partial_2 + \frac{\gamma}{N} \partial_4 \right) \right\}.$$

Now from TCM tables, or directly, the bivariate Faà di Bruno formula coefficients are those in the expression of μ'_{31} in terms of cumulants. Thus,

$$\mu'_{31} = \kappa_{31} + 3\kappa_{21}\kappa_{10} + \kappa_{30}\kappa_{01} + 3\kappa_{20}\kappa_{11} + 3\kappa_{11}\kappa_{10} + 3\kappa_{20}\kappa_{10}\kappa_{01} + \kappa_{10}^3 \kappa_{01}, \tag{A.7b}$$

leading to, in abbreviated notation, following the device of A.2.2,

$$E\left(\frac{L_1^3 L_4}{3! L_2^4} \right) \sim K_{31} + 3 K_{21} K_{10} + \ldots + K_{10}^3 K_{01}, \tag{A.8a}$$

where, for example,

$$K_{31} = \frac{N^3}{3! I^4} \int_0^1 \Phi^{N-1}\left(\frac{Ny}{I}\right) \left\{ \sum_0^\infty \frac{[1^3 4 2^r] y^r}{I^r r!} \right\} \frac{y^3}{3!} \, dy, \tag{A.8b}$$

$$K_{21} K_{10} = \frac{N^3}{3! I^4} \int_0^1 \Phi^{N-2}\left(\frac{Ny}{I}\right) \left\{ \sum_0^\infty \frac{[1^2 4 2^r] y^r}{I^r r!} \right\} \left\{ \sum_0^\infty \frac{[1 2^r] y^r}{I^r r!} \right\} \frac{y^3}{3!} \, dy, \tag{A.8c}$$

$$K_{20}K_{10}K_{01} = \frac{N}{3!I^4}\int_0^1 \Phi^{N-3}\left(\frac{Ny}{I}\right)\left\{\sum_0^\infty \frac{[1^2 2^r] y^r}{I^r r!}\right\}\left\{\sum_0^\infty \frac{[12^r] y^r}{I^r r!}\right\}$$

$$\times \left\{\sum_0^\infty \frac{[42^r] y^r}{I^r r!}\right\} \frac{y^3}{3!}\,dy, \tag{A.8d}$$

The ordered (in N^{-1}) terms in (A.8a) are found by determining the lowest-order term in each term. The rule to observe in ordering, i.e. for obtaining the lowest term M_s involved, is to subtract the number of factors in the cumulant terms in (A.7b), ignoring powers of κ_{10}, from the power of y in (A.8b), (A.8c) and (A.8d) enhanced by unity. For example,

κ_{31} relates to lowest-order $3 + 1 - 1 = 3$,
$\kappa_{21}\kappa_{10}$ relates to lowest-order $3 + 1 - 1 = 3$,
$\kappa_{20}\kappa_{11}$ relates to lowest-order $3 + 1 - 2 = 2$,
$\kappa_{11}\kappa_{10}$ relates to lowest-order $3 + 1 - 1 = 3$,
$\kappa_{10}^3\kappa_{01}$ relates to lowest-order $3 + 1 - 1 = 3$,
$\kappa_{20}\kappa_{10}\kappa_{01}$ relates to lowest-order $3 + 1 - 2 = 2$.

The reason for excepting powers of κ_{10} is that this corresponds in the integrands to a term

$$[1] + \frac{[12]y}{1!I} + \frac{[12^2]y^2}{2!I^2} + \ldots,$$

where, generally, $[1] = \sum \frac{\partial p_j}{\partial \theta} = \frac{\partial}{\partial \theta} \sum p_j = 0.$

Moreover, terms in (A.8b), (A.8c) and (A.8d) of the same order will arise from the basic products,

$$(1-y)^{N-1}y^r, (1-y)^{N-2}y^{r+1}, \ldots, (1-y)^{N-s}y^{r+s-1}.$$

A.2.4 Derivation of M_2^{-1} term in the bias

To consolidate the rather loose development so far, we now derive by this new approach the second-order factorial coefficient in $\mathrm{E}(\hat{\theta} - \theta)$. We have to consider four terms only of (A.3):

(a) $-L_1/L_2$

From (A.6) the contribution is

$$[12^2]/I^3 + 3[12][2^2]/I^4. \tag{A.9a}$$

(b) $-L_1^2 L_3/(2L_2^3)$

The "signature" expression is

$$\mu'_{21} = \kappa_{21} + 2\kappa_{10}\kappa_{11} + \kappa_{20}\kappa_{01} + \kappa_{10}^2\kappa_{01} \tag{A.9b}$$

and, for example,

$$K_{21} = \frac{N^2}{2I^3} \int_0^1 \left\{ \sum_0^\infty \frac{[2^r]y^r}{r!\,I^r} \right\}^{N-1} \sum_0^\infty \frac{[1^2 3 2^r]y^r}{r!\,I^r} \frac{y^2}{2!}\, dy.$$

Evidently, from the order rule, there are contributions from every term in (A.9b), and in particular the M_2^{-1} terms are

$$\left.\begin{aligned}
&\text{from } \kappa_{21}: \quad [1^2 3]/(2I^3)\\
&\text{from } 2\kappa_{10}\kappa_{11}: \quad 3[12][13]/I^4\\
&\text{from } 2\kappa_{20}\kappa_{01}: \quad 3[23]/(2I^3) + (3[1^2 2][3] + 3[2^2][3])/I^4\\
&\text{from } \kappa_{10}^2\kappa_{01}: \quad 6[12]^2[3]/I^5.
\end{aligned}\right\} \tag{A.9c}$$

(c) $L_1^3 L_4/(6L_2^4)$

The "signature" expression is now

$$\mu'_{31} = \kappa_{31} + 3\kappa_{21}\kappa_{10} + \underline{\kappa_{30}\kappa_{01}} + \underline{3\kappa_{20}\kappa_{11}} + 3\kappa_{11}\kappa_{10}^2 + \underline{3\kappa_{20}\kappa_{10}\kappa_{01}} + \kappa_{10}^3\kappa_{101},$$

the underscored terms being the only contributors, yielding

$$([1^3][4]/6I^4) + ([14]/(2I^3)) + (2[12][4]/I^4). \tag{A.9d}$$

(d) $-L_1^3 L_3^2/(2L_2^5)$

Here,

$$\mu'_{32} = \kappa_{32} + 3\kappa_{22}\kappa_{10} + 2\kappa_{31}\kappa_{01} + 3\kappa_{12}\kappa_{20} + 6\kappa_{21}\kappa_{11} + \kappa_{30}\kappa_{02}$$
$$+ 3\kappa_{12}\kappa_{10}^2 + 6\kappa_{21}\kappa_{10}\kappa_{01} + \underline{\kappa_{30}\kappa_{01}^2} + 6\kappa_{11}^2\kappa_{10} + 3\kappa_{20}\kappa_{02}\kappa_{10}$$
$$+ \underline{6\kappa_{20}\kappa_{11}\kappa_{01}} + \kappa_{02}\kappa_{10}^3 + 6\kappa_{11}\kappa_{10}\kappa_{01} + \underline{3\kappa_{20}\kappa_{10}\kappa_{01}} + \kappa_{10}\kappa_{01},$$

the contribution from the underscored terms being

$$([1^3][3]^2/(2I^5)) + (3[13][3]/I^4) + (15[12][3]^2/2I^5). \tag{A.9e}$$

(e) $(L_1^4/(4!L_2^4))(-L_5/L_2 + 10L_3 L_4/L_2^2 - 15L_3^3/L_2^3)$

The only terms contributing from the signature expressions for $\mu'_{41}, \mu'_{411}, \mu'_{43}$ are $3\kappa_{20}^2\kappa_{01}, 3\kappa_{200}\kappa_{010}\kappa_{001}$ and $3\kappa_{20}^2\kappa_{01}^3$ respectively, yielding

$$[5]/(8I^3) + 5[3][4]/(4I^4) + 15[3]^3/(8I^5). \tag{A.9f}$$

Adding the terms in (A.9c), (A.9d), (A.9e) and (A.9f) gives for the M_2^{-1} term in the bias,

$$\frac{1}{I^3}\left\{\frac{[1^2 3]}{2} + \frac{[14]}{2} + \frac{[5]}{8} + \frac{3[23]}{2} + [12^2]\right\} +$$

$$+ \frac{1}{I^4}\left\{3[12][13] + \frac{3}{2}[1^2 2][3] + 3[2^2][3] + \frac{[1^3][4]}{6}\right.$$

$$\left. + 2[12][4] + 3[13][3] + \frac{5}{4}[3][4] + 3[12][2^2]\right\}$$

$$+ \frac{1}{I^5}\left\{6[12]^2[3] + [1^3][3]^2 + \frac{15}{8}[3]^3 + 6[12][3]^2\right\},$$

which agrees with $a_3^{(1)}, a_4^{(1)}, a_5^{(1)}$ of (2.30a).

A.3 The general case

From a theoretical viewpoint there are three stages in evaluating the coefficient of M_s^{-1} in the bias:

(a) The value of s determines that terms in (A.3) must be scanned as far as the term L_1^{2s}. It will be necessary to generate the Lagrange series to this term.

(b) The set of multivariate formulae for μ' in terms of the κ's, including all derivatives corresponding to the structure of the L's in (A.3), must be set up.

(c) The evaluation of integrals corresponding to the admissible κ-products in the μ''s must be formulated.

With regard to (b), the multivariate derivatives of functions of functions may be set up in two ways at least. First, the method described in the TCM tables is available; this relies on Kendall's (1940) operational technique. For example, from

$$\mu_4' = \kappa_4 + 4\kappa_3\kappa_1 + 3\kappa_2^2 + 6\kappa_2\kappa_1^2 + \kappa_1^4,$$

written in the form

$$\mu'(r^4) = \kappa(r^4) + 4\kappa(r^3)\kappa(r) + \ldots,$$

we derive, using the operator $s\partial/\partial r$,

$$\mu'(r^3 s) = \kappa(r^3 s) + 3\kappa(r^2 s)\kappa(r) + \kappa(r^3)\kappa(s) + \ldots,$$

corresponding to

$$\mu'_{31} = \kappa_{31} + 3\kappa_{21}\kappa_{10} + \kappa_{30}\kappa_{01} + 3\kappa_{20}\kappa_{11} + 3\kappa_{11}\kappa_{10}^2 + 3\kappa_{20}\kappa_{10}\kappa_{01} + \kappa_{10}^3\kappa_{01},$$

which in turn yields the formula for a fourth derivative of a composite function,

$$\frac{\partial^4}{\partial x^3 \partial y} F(f(x,y)) = F_1 f_{31} + 3F_2 f_{21} f_{10} + F_2 f_{30} f_{01} + 3F_2 f_{20} f_{11}$$

$$+ 3F_3 f_{11} f_{10}^2 + 3F_3 f_{20} f_{10} f_{01} + F_4 f_{10}^3 f_{01},$$

where

$$F_s = \frac{\partial^s F}{\partial f^s}$$

$$f_{ij} = \frac{\partial^{i+j}}{\partial x^i \partial y^j} f(x,y).$$

A program (described in TCM) was set up by Dr Kratky and Dr Reinfelds using the AMTRAN language, and this led to the setting up of formulae up to tenth derivatives.

Another possibility would be to utilize Good's formula (1961, pp. 540–2), but we do not have information on this approach.

A third possibility would be to use the basic unitary formulae for μ' in terms of κ's (this idea arose out of a conversation between the present authors and Juris Reinfelds at the Vienna ISI conference (1973)). An illustration will suffice. There is the formula

$$\begin{aligned}\mu'_{1111} =\ & \kappa_{1111} + \kappa_{0111}\kappa_{1000} + \kappa_{1110}\kappa_{0001} + \kappa_{1011}\kappa_{0100} + \kappa_{1101}\kappa_{0010} \\ & + \kappa_{0101}\kappa_{1010} + \kappa_{1100}\kappa_{0011} + \kappa_{1001}\kappa_{0110} + \kappa_{0110}\kappa_{1000}\kappa_{0001} \\ & + \kappa_{0011}\kappa_{1000}\kappa_{0100} + \kappa_{1010}\kappa_{0001}\kappa_{0100} + \kappa_{0101}\kappa_{1000}\kappa_{0010} \\ & + \kappa_{1100}\kappa_{0001}\kappa_{0010} + \kappa_{0101}\kappa_{1000}\kappa_{0010} + \kappa_{1000}\kappa_{0001}\kappa_{0100}\kappa_{0010} \end{aligned} \quad (A.10)$$

in which all the coefficients are unity, and the vector unity on the left is arranged in all possible ways over one, two, . . . , four factors (*order of the factors being ignored*) *on the right.* From this, by summation of corresponding subscripts, any fourth-order derivative can be set up. For example, $\mu'_{31} \equiv \mu'_{(1+1+1),1}$ (or for that matter any corresponding sets of three) and so

$$\mu'_{31} = \kappa_{31} + 3\kappa_{21}\kappa_{10} + \ldots.$$

Of course, the number of terms in formulae similar to (A.10) would be large for higher derivatives; for μ'_1, a vector with ten components, there are 115975 distinct terms, and these, when grouped to give a result for a vector with diminished components, might present a problem in pattern recognition.

For (c) it is a matter of evaluating a Beta-function type integral, corresponding to a term such as $k_{100}^{s_1}\ldots k_{010}^{s_2}\ldots k_{120}^{t_1}\ldots$; the integrand is expanded in descending powers of $(1-y)$, only terms of the same order in the sample size being selected.

The application to higher moments such as the variance is carried through by similar methods, the partition patterns of the terms being sustained, provided the forms worked with are $\mathrm{E}(\hat{\theta} - \theta)^s, s = 2, 3, \ldots$, followed by suitable correction formulae.

Does the approach have advantages over the technique of Chapters 2 and 3? The expansions, in the general case, will be unexpectedly complicated, no matter how derived. For example, by the factorial sample-size approach (which is almost certain to contain fewer terms than the non-factorial) for the bias, the numbers of terms in $M_1^{-1}, M_2^{-1}, M_3^{-1}$ are 2, 17, 112 respectively; a fairly careful attempt at the M_4^{-1} terms indicates that there will be 600 terms approximately, which, though large by some standards, is not a serious problem for a digital approach. The main problem is to generate the $\mu - \kappa$ formulae, to analyse the structure of each term as to its weight as a contributor to $M_1^{-1}, M_2^{-2}, \ldots$, and finally to interpret it as a Beta-integral.

A.4 Factorial series for bias and variance up to N^{-3}

For convenience, we list these expressions in terms of powers of I^{-1}.

A.4.1 Bias

$$E(\hat{\theta} - \theta) \sim \sum_{s=1}^{\infty} \frac{f_s^{(1)}}{(N+1)(N+2)\ldots(N+s)}, \quad (A.11)$$

where

(i) $f_1^{(1)} = ([12] + \tfrac{1}{2}[3])/I^2$;

(ii) $f_2^{(1)} = a_3^{(1)}/I^3 + a_4^{(1)}/I^4 + a_5^{(1)}/I^5$,

where

$8a_3^{(1)} = 4[1^23] + 8[12^2] + 4[14] + 12[23] + [5]$,

$12a_4^{(1)} = 2[4][1^3] + 18[3][1^22] + 15[3][4] + 36[3][2^2] + 36[3][13]$
$\qquad + 36[12][13] + 36[12][2^2] + 24[4][12]$,

$8a_5^{(1)} = 4[3]^2[1^3] + 15[3]^3 + 48[3][12]^2 + 60[3]^2[12]$;

(iii) $f_3^{(1)} = \sum_{s=4}^{8} B_s^{(1)}/I^s$,

where

$48B_4^{(1)} = 8[1^34] + 72[1^223] + 48[12^3] + 12[1^25] + 96[124]$
$\qquad + 144[2^23] + 72[13^2] + 6[16] + 30[25] + 60[34] + [7]$,

$48B_5^{(1)} = 3[3]\{16[1^33] + 48[1^22^2] + 40[1^24] + 240[123] + 80[2^3]$
$\qquad + 30[15] + 120[24] + 90[3^2] + 7[6]\} + 8[4]\{4[1^32] + 15[1^23]$
$\qquad + 30[12^2] + 15[14] + 45[23]\} + [5]\{2[1^4] + 60[1^22] + 90[13]$
$\qquad + 90[2^2] + 35[4]\} + 4[1^3]\{2[15] + 8[24] + 6[3^2] + [6]\}$
$\qquad + 12[12]\{8[1^24] + 48[123] + 16[2^3] + 10[15] + 40[24] + 30[3^2]$
$\qquad + 3[6]\} + 96[1^22][14] + 288[12^2][13] + 288[1^22][23]$
$\qquad + 144[1^23][2^2] + 144[1^23][13] + 288[12^2][2^2] + 240[13][14]$
$\qquad + 240[14][2^2] + 720[23][13] + 720[2^2][23]$,

$24B_6^{(1)} = [3]\{10[4][1^4] + 360[1^22][2^2] + 105[3][5] + 140[4]^2$

$$
\begin{aligned}
&+ 360[4][1^22] + 540[3][12^2] + 60[3][1^32] + 40[1^3][14] \\
&+ 360[1^22][13] + 360[1^23][12] + 120[1^3][23] + 270[3][1^23] \\
&+ 30[5][1^3] + 540[13]^2 + 1080[13][2^2] + 540[2^2]^2 + 315[3][14] \\
&+ 630[4][13] + 945[3][23]\} + 2[4]\{20[1^3][13] + 120[1^22][12] \\
&+ 20[1^3][2^2] + 10[4][1^3] + 315[3][2^2]\} + [12]\{20[5][1^3] \\
&+ 240[14][12] + 360[13]^2 + 720[23][12] + 720[2^2][13] + 360[2^2]^2 \\
&+ 720[3][14] + 720[4][13] + 2160[3][23] + 720[4][2^2] \\
&+ 180[5][12] + 315[5][3] + 210[4]^2 + 720[3][12^2]\},
\end{aligned}
$$

$$
\begin{aligned}
8B_7^{(1)} =\ & [3]^2\{5[3][1^4] + 60[1^3][13] + 70[4][1^3] + 210[3][1^22] \\
&+ 360[1^22][12] + 60[1^3][2^2] + 1260[12][13] + 1260[2^2][12] \\
&+ 420[3][13] + 420[3][2^2] + 480[4][12] + 210[4][3]\} + 8[3] \\
&\{10[4][1^3][12] + 90[13][12]^2 + 90[2^2][12]^2 + 105[4][12]^2\} \\
&+ 160[4][12]^3,
\end{aligned}
$$

$$
\begin{aligned}
16B_8^{(1)} =\ & [3]^2\{140[3]^2[1^3] + 280[3][1^3][12] + 1680[12]^3 \\
&+ 3360[3][12]^2 + 1890[3]^3[12] + 315[3]^3\}.
\end{aligned}
$$

A.4.2 Second non-central moment

$$
\mathrm{E}(\hat{\theta} - \theta)^2 \sim \sum_{1}^{\infty} \frac{p_s}{(N+1)(N+2)\ldots(N+s)}, \tag{A.12}
$$

where

$$
\begin{aligned}
p_1 =\ & 1/I, \\
p_2 =\ & (3[2^2] + 3[13] + 2[1^22] + [4])/I^3 + (6[12]^2 + 12[12][3] \\
& + (15/4)[3]^2 + [3][1^3])/I^4, \\
p_3 =\ & a/I^4 + b/I^5 + c/I^6 + d/I^7,
\end{aligned}
$$

with

$$
\begin{aligned}
4a =\ & 4[1^33] + 12[1^22^2] + 16[2^3] + 48[123] + 8[1^24] + 20[24] \\
& + 15[3^2] + 5[15] + [6],
\end{aligned}
$$

$$
\begin{aligned}
12b =\ & 4[4][1^4] + 48[3][1^32] + 144[1^22][13] + 144[1^23][12] \\
& + 48[1^3][23] + 144[1^22][2^2] + 288[12^2][12] + 360[3][12^2] \\
& + 180[3][1^23] + 120[4][1^22] + 16[1^3][14] + 10[5][1^3] \\
& + 240[12][14] + 720[23][12] + 360[13][2^2] + 540[3][23] \\
& + 180[4][2^2] + 180[2^2]^2 + 90[5][12] + 180[3][14] + 180[13]^2 \\
& + 180[4][13] + (105/2)[5][3] + 35[4]^2,
\end{aligned}
$$

$$
12c = 15[3]^2[1^4] + 720[3][1^22][12] + 120[3][1^3][2^2] +
$$

$$+ 540[3]^2[1^22] + 120[3][1^3][13] + 80[4][1^3][12] + 120[4][3][1^3]$$
$$+ 720[13][12]^2 + 2160[3][12][2^2] + 2160[12][13][3]$$
$$+ 720[4][12]^2 + 945[3]^2[2^2] + 720[2^2][12]^2 + 1260[4][3][12]$$
$$+ 945[3]^2[13] + 420[4][3]^2,$$

$$4d = 120[3]^2[1^3][12] + 70[3]^3[1^3] + 480[3][12]^3 + 1260[3]^2[12]^2$$
$$+ 840[3]^3[12] + 315[3]^4/2.$$

REFERENCES

Comtet, L. (1974). *Advanced Combinatorics.* Reidel Publishing Co., Boston.

Good, I. J. (1961). "The multivariate saddlepoint method and chi-squared for the multinomial distribution". *Ann. Math. Statist.,* **32**, 535–48.

Kendall, M. G. (1940). "The derivation of multivariate sampling formulae from univariate formulae by symbolic operation". *Ann. Eugen.,* London, **10**, 392–402.

McMahon, J. (1894). "On the general term in the reversion of series". *Bull. Amer. Math. Soc.,* **3**, 170–2.

Riordan, J. (1946). "Derivatives of composite functions". *Bull. Amer. Math. Soc.,* **52**, 664–7.

APPENDIX B

SUMMATION OF ASYMPTOTIC SERIES

B.1 Introduction

Examples have been given of asymptotic series developments for mle, either in the sample size N or another parameter (see, for example, the estimator for θ in the L.S.D. (5.2), and the estimators for ρ and a in the Ga.D. (5.4)). The series for $E\hat{\theta}$ in descending N (Tables 5.1A, 5.1B, and 5.1C) show irregular sign patterns and divergent tendencies, and require summation techniques especially for θ small and $N < 8$; a similar remark applies to the higher moments. By contrast, the series for the moments of $\hat{\rho}$ in sampling from the Ga.D. (Table 5.11) show a regular sign pattern and behave approximately as geometric progressions. But another estimator of ρ, the moment estimator, has quite different properties. Defining $\bar{\rho} = m_1^2/m_2$ (m_1 the sample mean, m_2 the sample second central moment), we have, when $a = \rho = 1$, the expression for the bias

$$E\bar{\rho} \sim 1 + \frac{6}{N} - \frac{38}{N^2} + \frac{2358}{N^3} - \frac{217318}{N^4} + \frac{30753942}{N^5} - \frac{6052130034}{N^6},$$
(B.1)

showing an evident divergent tendency. In fact, the ratios of successive coefficients are 6, 6·3, 62, 92, 142, and 196 approximately.

B.2 Summation algorithms

B.2.1 Let the series to be summed be

$$S(N) \sim e_0 + e_1/N + \ldots + e_r/N^r, \quad (e_0 \neq 0).$$
(B.2)

In cases of non-regular estimation (Chapter 4) r will generally be small (≤ 6) and firm assessments of moments using summation devices will at best be tentative. We shall assume that $r \geq 8$, although if the statistic under consideration is simply structured and a low-order moment is under consideration, the first few terms may display the dominant trend. One would expect, for example, aberrant behaviour in the first few coefficients of a series in N^{-1} for the fourth (or higher) central moment of a statistic.

B.2.2 Padé approximants

A general form is

$$P[s/t] = \frac{n_0 + n_1/N + \ldots + n_s/N^s}{d_0 + d_1/N + \ldots + d_t/N^t} \quad (n_0 = e_0,\ d_0 = 1), \tag{B.3}$$

where the coefficients are determined from the asymptotic equivalence of (B.2) and (B.3), and

$$S(N) - P[s/t] = O(N^{-s-t}). \tag{B.4}$$

For example, for (B.1),

$$\left.\begin{aligned}
P[1/0] &= 1 + 6/N \quad P[0/1] = 1/(1 - 6/N) \\
P[2/1] &= \frac{19 + 1293/N + 6352/N^2}{19 + 1179/N} \\
P[1/2] &= \frac{37 + 1515/N}{37 + 1293/N - 6352/N^2} \\
P[2/2] &= \frac{397 + 40329/N + 296906/N^2}{397 + 37947/N + 84310/N^2}
\end{aligned}\right\} \tag{B.5}$$

Padé approximants were introduced in the latter part of the nineteenth century and were mainly of theoretical interest. Recently, however, they have been extensively used and researched by theoretical physicists, finding application in the solution of differential equations, linear integral equations and many other situations (Baker and Gammel (1970)). The change in attitude to the subject can be seen in the references to it in the classical books of Perron (1929) and Wall (1948), compared with the emphasis to be found in Baker (1975), who is one of the leading present-day writers.

The development of digital computing facilities in the last half-century has doubtless been mainly responsible for this resurgence of interest, bringing within reach higher-order terms that are impossible to obtain by purely mechanical devices.

B.2.3 Extrapolation of series

If a series shows a regular sign pattern and the coefficients are simply structured ($|e_s/e_{s-1}| \to k$ for large s, for example) then a mimicking series can be constructed as an improved approximator (see Gaunt and Guttmann (1975)). For example, the mle for ρ (Table 5.11) when $\rho = 25, a = 1$, has relative bias

$$E(\hat{\rho} - \rho)/\rho \sim 2 \cdot 97/N + 8 \cdot 92/N^2 + 26 \cdot 8/N^3$$
$$+ 80 \cdot 3/N^4 + 241/N^5 + 723/N^6 \tag{B.6}$$
$$= b(N) \text{ say},$$

the entries being rounded-off for convenience. Clearly the series is approximately a geometric progression with ratio nearly three. A better approximant is ($N>3$)

$$E(\hat{\rho} - \rho)/\rho \sim b(N) - 723/N^6 + 723/(N^6 - 3N^5). \tag{B.7}$$

The validity of this extrapolant could be tested by Monte Carlo simulations, although the effect of the refinement of the added term might be difficult to pin down (see references 2 and 3 following **5.4**).

B.2.4   Borel–Padé algorithms

The basic idea springs from Émile Borel's (1928) work on divergent series. For the expansion

$$f(x) = \sum_{s=0}^{\infty} e_s x^s, \tag{B.8}$$

for which $|e_s|$ has a large order of magnitude (such as $(ks)!$ with $k>1$), we consider the modified series

$$f(x) \sim \int_a^b \left\{ \sum_{s=0}^{\infty} f_s(xy)^s \right\} d\sigma(y), \tag{B.9}$$

where $\sigma(\cdot)$ is a suitable non-decreasing function whose moments exist. The integral is to be evaluated by using the partial fraction form of Padé approximants to the integrand. It is customary to consider (Graffi et al. (1971))

$$f(x) \sim (1/m) \int_0^{\infty} \{\exp(-y^{1/m})\} f^*(xy) y^{-1+1/m} \, dy, \tag{B.10}$$

$$(m>0)$$

where $f^*(z) = \sum_{s=0}^{\infty} f_s z^s / \Gamma(sm+1)$.

B.2.5   Modified Borel–Padé algorithms

Another version of the algorithm has been studied by the authors (Shenton and Bowman (1976), Bowman and Shenton (1976)). The one-component model consists of approximating a series $S(N)$ by means of

$$F_r(N; \sigma) = N\phi_{r-1}(N) + \psi_r(N)R(N), \tag{B.11}$$

where $\phi_{r-1}(\cdot)$, $\psi_r(\cdot)$ are polynomials in N of degrees $r-1$ and r respectively and

$$R(N) = \int_a^b \frac{d\sigma(t)}{N+t},$$

where $\sigma(\cdot)$ is a non-decreasing function whose moments exist. It is shown that if

$$R(N) = \frac{a_0}{N+b_1} - \frac{a_1}{N+b_2} - \ldots$$

then, under certain conditions,

$$F_r(N;\sigma) = N \sum_{s=0}^{r} \frac{(\omega_s(N)R(N) - \psi_s(N))}{a_0 a_1 \ldots a_s} \omega_s^*, \tag{B.13}$$

where the sth convergent of (B.12) is $\psi_s(N)/\omega_s(N)$ ($\psi_0 = 0$, $\psi_1 = a_0$; $\omega_0 = 1$, $\omega_1 = N + b_1$), and ω_s^* is an interfacing function involving the coefficients of $R(N)$ and the coefficients $e_0, e_1, \ldots$ of $S(N)$. To derive ω_s^*, a power of N in $\omega_s(N)$ is replaced by the coefficient of the corresponding power of N^{-1} in $S(N)$; for example, $\omega_0^* = e_0$, $\omega_1^* = b_1 e_0 + e_1$.

The 2-component version similarly consists of

$$F_r(N) = N\pi_*^*(N) + \left(B_0^{(r)} - B_2^{(r)}\frac{N}{2!} + B_4^{(r)}\frac{N^2}{4!} - \cdots\right)\omega_{a-1}(N)$$
$$+ \left(B_1^{(r)} - B_3^{(r)}\frac{N}{3!} + B_5^{(r)}\frac{N^2}{5!} - \cdots\right)\omega_a(N), \tag{B.14}$$

involving r parameters $B_0^{(r)}, B_1^{(r)}, \ldots$, and

where $\quad \omega_i(N) = \int_0^\infty \frac{e^{-t} t^i dt}{1 + t^2/N} \quad (i = a-1, a; a > 0)$

and for which

$$S(N) - F_r(N) = O(N^{1-r}).$$

The polynomial $\pi_*^*(\cdot)$ in N is introduced to compensate for the redundant coefficients of $N, N^2, \ldots$, introduced in the polynomial multipliers of ω_{a-1} and ω_a. The evaluation of the parameters $B_0^{(r)}, B_1^{(r)}, \ldots$ involves a simple recursive formula in terms of the coefficients $e_0, e_1, \ldots$.

B.3 Illustrations

B.3.1 $\hat{\theta}$ in the logarithmic series distribution (5.2)

The coefficients of the first nine powers of N^{-1} in $E\,\hat{\theta}$ when $\theta = 0.1$ (see Table 5.1C) are $-0.1558556, 0.2107310, -0.2356316, 0.1596725, 0.1647786, -0.9802963, 2.461617, -3.750214$ and -0.6570077, the zero order term being 0.1. The series requires a summation technique for $2 \le N < 8$. Padé approximants (Table B.1) produce results in good agreement with the correct values (the entries in Table 5.3 are thought to be correct to at least five decimal digits). Comparisons are also shown for $\mu_2(\hat{\theta})$.

To illustrate the one-component Borel model (B.11) we have used the periodic continued fraction

Table B1
Padé approximants $P[s/s]$ to $E\hat{\theta}$ and $\mu_2(\hat{\theta})$ (L.S.D. parameter)

s		Sample size, N			
		2	3	4	5
2	a	0·0547148	0·0647536	0·0711873	0·0756495
	b	0·0282153	0·0255724	0·0228335	0·0204288
3	a	0·0546310	0·0647201	0·0711716	0·0756412
	b	0·0270819	0·0252538	0·0227173	0·0203786
4	a	0·0546412	0·0647228	0·0711725	0·0756416
	b	0·0274283	0·0253192	0·0227345	0·0203842
5	a	0·0546409	0·0647227	0·0711724	0·0756416
	b	—	—	—	—
Correct	a	0·05464	0·06472	0·07117	0·07564
value	b	0·02743	0·02532	0·02273	0·02038

(a refers to $E\hat{\theta}$, b to $\mu_2(\hat{\theta})$. The correct values are taken from Table 5.3.)

$$R(N) = \frac{1}{N+1} - \frac{1}{N+2} - \frac{1}{N+2} - \cdots$$

in the series for $E\hat{\theta}$ when $\theta = 0\cdot1$. It turns out that the ninth approximants when $N = 2, 3, 4, 5$ are 0·054641, 0·064723, 0·071172, and 0·075642, in excellent agreement with the corresponding Padé assessments and the true values.

.4 Remarks

This brief account of summation techniques scarcely does justice to a developing subject which attempts to discover properties of functions for which most of the information lies in a Taylor-series development. No mention has been made of methods of treating series known to have singularities (with unspecified residues), and the effect of these on the sum evaluation. There is also the ever-present problem of loss of accuracy in complex digital manipulations; and before results are finalized different algorithms should be compared which have been implemented on different machines.

REFERENCES

Baker, G. A. Jr. (1975). *Essentials of Padé Approximants*. Academic Press, New York.

Baker, G. A. Jr., and Gammel, J. L. (1970). *The Padé Approximant in Theoretical Physics*. Academic Press, New York.

3. Borel, É. (1928). *Leçons sur les séries divergentes*. Gauthier-Villars, Paris
4. Bowman, K. O. and Shenton, L. R. (1976). "A new algorithm for summing divergent series: Part 2: A two-component Borel summability model". *J. Computational and Applied Math.*, **2**, No. 4.
5. Gaunt, D. S. and Guttmann, A. J. (1975). "Asymptotic analysis of coefficients", Chapter 4 in *Phase Transitions and Critical Phenomena*, **3**, C. Domb and M. S. Green, editors. Academic Press, New York and London.
6. Graffi, S., Grecchi, V. and Turchetti, G. (1971). "Summation methods for the perturbation series of the generalized anharmonic oscillator". *Il Nuovo Cimento*, **4B**, No. 2, 313–40.
7. Perron, O. (1929). *Die Lehre von den Kettenbrüchen*. B. G. Teubner, Leipzig and Berlin.
8. Shenton, L. R. and Bowman, K. O. (1976). "A new algorithm for summing divergent series; Part 1: Basic theory and illustrations". *J. Computational and Applied Math.*, **2**, No. 3, 151–67.
9. Wall, H. S. (1948). *Analytic Theory of Continued Fractions*. D. Van Nostrand, New York.

INDEX

Adjusted order of magnitude method, 38–41, 77
Asymptotic(s):
  bias, 5
  covariance matrix, 10, 11, 93
  efficiency, 10, 160
  expansions, 26
  first-order, 11, 13
  joint efficiency, 10
  moments, 25
  normality, 6, 7, 9
  series, 21, 23, 24, 25, 26, 27, 29, 30, 31, 130–3, 141–2, 151, 153–4, 155–6, 166, 167, 174–6, 177
  summation, Appendix B
  variance, 12, 14, 46–7, 53–4, 56, 57, 58–9, 59–60, 61

Bernoulli numbers, 21, 24, 27
Bias (mle(s)):
  single parameter, first-order, 38, 68
    to third order, 44–6, 51–3
    factorial series, 174
  single parameter, special cases:
    linear probability, 56
    $N(\theta, 1)$, 57
    $N(0, \theta)$, 58–9
    Poisson, 59–60
    Fisher's Nile problem, 60
  two parameters, first-order, 69
  multiparameter, first-order, 69
    second-order, 70–1
    special cases:
      one estimator sample mean, 75–6
      negative binomial, 77
      $N(\mu, \sigma^2)$, 69
      bivariate normal, 69–70
  of estimator, 5
Bienaymé–Chebyshev inequality, 3
Binomial distribution, 25
  (negative), 10, 74–7, 80–90, 162
Bivariate normal, 69–70
Borel–Padé algorithms, 179
  modified algorithms, 179–81

Calculus of variations, 8
Categorized data, 2

Cauchy distribution, 18–21
Central limit theorem, 7
Central moments, 46–9, 53–5, 56–61
Consistency, 3, 7
Correlation, 108
Covariance (mle(s)):
  multiple parameter, first-order, 66
    second-order, 67
  special cases:
    orthogonal parameters, 72–3
    two parameters with one the mean, 73
    negative binomial, 74–5
Cramér–Rao bound, 8, 25
Cumulants and Faà di Bruno's formula, 169–74
c-variate Taylor approach, 41–4

Derivatives, 41, 64, 129
  of a function of a function, 172–3
  of a probability function, 51–6
Distribution of precipitation and the gamma density, 149
Divergent series, 23, Appendix B

Efficiency, 7, 10
Estimation, 9, 22–3
Expectation of linear forms, 35–6
Exponential regression, 3, 121–5
Extrapolation of series, 178–9

Faà di Bruno's formula, 14, 130, 142, 150, 169
Factorial series for bias (mle), 174–5
  for second moment, 175–6

Gamma density, two parameters, 149–57
  three parameters, 110–13
Generalized variance, 9–10

Hermite distribution, 113–21

Information, $I(\theta)$, 6, 34–5
Inverse efficiency law, 81, 108–10

Jackknife, 5
Joint efficiency, 10

Kronecker function, 20
Kurtosis, 17, 30, 123–5, 138–9, 149, 152–3
  single parameter (mle), N^{-1}, N^{-2} terms, 49, 55
  special cases, $N(\theta, 1)$, 57–8

Lagrange's equation (expansion), 19, 34, 36, 37, 122, 123, 167
Likelihood equation, 6, 7, 9, 18, 34, 63
  function, 1, 17, 34
Logarithmic derivatives, 44–50
Logarithmic series distribution, 126–43
  and Padé approximants, 180–1

Maximum likelihood estimator(s), Chapter 1
McMahon's analysis of Lagrange's expansion, 167
Method of moments, 158–61
Mimic series, 178
Moment estimators, 158–61, 162–3
Moments into cumulants, 169–74
MVU, 9

Negative binomial distribution, 74–7, 80–91
Neyman Type A distribution, 91–108
Nile, Fisher's problem, 60–2
Non-regular estimation, Chapter 4
Normal distribution, 4, 6–7, 9, 57–9
  probability integral, 24
  mixtures, second-order terms, 161
Notations:
  Γ_s, 34
  $(n_j \Gamma^j)$, 34
  $(1^a 2^b \ldots)$, 51
  $[1^a 2^b \ldots]$, 44
  $(p \Gamma^{r_1} \Gamma^{r_2} \ldots)$, 44
  $I(\theta)$, 6, 34–5
  summatory, 34, 63
Number of terms in bias and covariance, 71, 75, 79

Order of magnitude of linear forms, 36

  of magnitude of sampling moments, 42
Orthogonal parameters, 72, 74
  polynomials, 113, 159

Padé approximants, 177–8
Partition function, 24–5
Partitions and Lagrange's equation, 166–74
Poisson distribution, 6, 13, 59, 81, 108
Pólya–Aeppli distribution, 108–10
Precipitation and the gamma density, 149

Regression model, 3, 121–5

Sampling moments (contributing terms), 42
Schwarz inequality, 8
Skewness, single parameter (mle), 49, 55
  special case, $N(\theta, 1)$, 57–8
  illustrations, 17, 30, 123–5, 138–9, 149, 152–3
Standard deviation, $E\sqrt{m_2}$ in normal sampling, 26
Stirling's expansion, 24
Sufficiency, 12
Summation of series, Appendix B

Taylor series, 22, 41, 63–4, 129, 143–4

Unbiased, 5
  estimators for two-parameter gamma density, 155–6
Unitary formulae for moments into cumulants, 173

Variance, single parameter (mle), 46–7, 53
  multiparameter (mles), 66–7
  generalized, 9–10, 93

Zero-truncated binomial distribution, 143–9
  Poisson distribution, 13–17

AUTHORS CITED

Aitken, A.C., 8, 9
Anscombe, F.J., 6, 74, 90, 93
Antle, C.E., 23
Bain, L.J., 23
Baker, George A. Jr., 178
Barger, G.L., 149
Bartlett, M.S., 6, 38
Barton, D.E., 93
Blischke, W.R., 156
Borel, É., 179
Bowman, K.O., 27, 29, 35, 38, 71, 77, 93, 110, 113, 120, 126, 145, 150, 152, 179
Brillinger, D.R., 5
Comtet, L., 167
Cox, D.R., 1, 35, 121, 122, 124
Craig, C.C., 30
Cramér, H., 1, 6, 7, 8
Crutcher, H.L., 149
D'Agostino, R., 155
David, F.N., 30
De Bruijn, 25
Derksen, J.B.D., 30
Durand, D., 150
Edwards, A.W.F., 1
Erdélyi, A., 25
Evans, D.A., 90
Faà di Bruno, 14, 130, 142, 144, 150
Feller, W., 1
Fisher, R.A., 2, 3, 7, 10, 18, 20, 27, 30, 35, 60, 90
Freeman, H., 8
Fryer, J.G., 161
Gart, J.J., 145
Gammel, John L., 178
Gaunt, A.B., 178
Geary, R.C., 10
Godwin, H.J., 3
Good, I.J., 63, 169, 173
Graffi, S., 179
Gray, H.L., 6
Grecchi, V., 179
Greenwood, J.A., 150
Gurland, J., 10
Guttman, A.J., 178
Haldane, J.B.S., 3, 35, 38, 57, 90
Hardy, G.H., 24, 25
Hartley, H.O., 154

Hinkley, D.V., 1
Huzurbazar, V.S., 5, 72
Jeffreys, H., 25, 72
Johnson, N.L., 30, 158
Kamat, A.R., 126, 128
Katti, S.K., 10
Kemp, A.W., 120
Kemp, C.D., 120
Kendall, M.G., 1, 3, 5, 6, 9, 11, 12, 57, 172
Kotz, S., 158
Kruskal, M.D., 25
Mason, T.J., 161
McKay, G.F., 149
McMahon, J., 167
Mooley, Diwakar, 149
Myers, R.H., 90
Norden, R.H., 1, 6
Patel, Y.C., 114, 119, 120
Patil, G.P., 126, 128, 140
Pearson, E.S., 12, 30, 31, 154, 155
Perron, O., 178
Quenouille, M.H., 5
Ramanujan, S., 24
Rao, C.R., 1, 6, 7, 8, 114, 120
Riordan, J., 128, 169
Robertson, C.A., 161
Ryzin, J. Van, 161
Schucany, W.R., 6
Sclove, S.L., 161
Sheehan, D.M., 27, 90
Shenton, L.R., 8, 27, 29, 35, 38, 71, 77, 90, 93, 110, 113, 120, 126, 145, 150, 152, 179
Silverstone, H., 8
Skees, P., 126
Smith, C.A.B., 5
Smith, Sheila M., 3, 35, 38, 57
Snell, E.J., 35, 121, 122, 124
Sprott, D.A., 73
Stuart, A., 1, 3, 5, 6, 9, 11, 12, 57
Thiele, T.N., 30
Thom, H.C.S., 149, 150
Thomas, D.G., 145
Tschuprow, Ap.A., 30
Tukey, J.W., 5
Turchetti, G., 179
Vestal, I.B., 149

Wall, H.S., 178
Wallace, D.L., 30
Wallington, P.A., 38, 69, 70
Wani, J.K., 126, 128, 140

Watson, G.N., 19, 24
Whittaker, E.T., 19, 24
Wilks, S.S., 9